CONSTRUCTION
MATERIALS FOR CIVIL ENGINEERING

Second Edition

Errol van Amsterdam

JUTA

I0044182

Construction Materials for Civil Engineering

First published 2000
Reprinted 2012
Second edition 2013
Reprinted 2013
Juta and Company Ltd
PO Box 14373, Lansdowne, 7779, Cape Town, South Africa

© 2013 Juta & Company Ltd

ISBN 978 0 70219 757 4

All rights reserved. No part of this publication may be reproduced or transmitted in any form or by any means, electronic or mechanical, including photocopying, recording, or any information storage or retrieval system, without prior permission in writing from the publisher. Subject to any applicable licensing terms and conditions in the case of electronically supplied publications, a person may engage in fair dealing with a copy of this publication for his or her personal or private use, or his or her research or private study. See Section 12(1)(a) of the Copyright Act 98 of 1978.

Project Manager: Karen Froneman
Editor: Jenny de Wet
Proofreader: Lee-Ann Ashcroft
Illustrator: Rassie Erasmus
Indexer: Daphne Burger
Cover designer: Marius Roux, MR Design
Typesetter: Trace Digital Services

Print Administration by DJE Flexible Solutions

The author and the publisher believe on the strength of due diligence exercised that this work does not contain any material that is the subject of copyright held by another person. In the alternative, they believe that any protected pre-existing material that may be comprised in it has been used with appropriate authority or has been used in circumstances that make such use permissible under the law.

The author and publishers wish to acknowledge the following persons and institutions for their contributions and permission to reproduce material in this book.
- This revised edition has tried as far as possible to stay abreast of the latest changes in material specifications and standard codes. We therefore acknowledge the South African National Standards (SANS) organisation and the International Organization for Standardization (ISO) for providing all the relevant information.
- Figure 4.7: Photographer Luciana Baptista Cohen and Warren Lee Cohen bread oven maker and workshop leader.
- Figures 3.2, 3.4, 4.5, 5.1: MediaClubSouthAfrica.com
- Figure 6.1: Natural Stone Warehouse

A note to the student

As with the first edition, this book does not aim to be a comprehensive reference book on construction materials, but is rather a thorough introduction to the subject covering and complementing the content acquired in the first year of the civil engineering diploma.

Civil Engineering has a wide variety of specialist fields and this book does not attempt to address any one in great detail but rather aims to provide the tools to build your knowledge and understanding. The book consists of six chapters, including a new chapter covering the internal construction materials used in finishes. Each chapter has been structured to test your understanding through self-evaluation questions as well as practical activities helping you to relate to the concepts explained in the book.

The language is kept simple and conversational in order to explain technical terminology and concepts.

Fully worked-out examples are provided within each section and practice questions are included in the self-evaluation sections. The answers to the self-evaluation questions appear at the end of each unit.

We have also established a link between this subject and civil engineering drawings to further demonstrate the various aspects of engineering in practice.

You will notice that the concept of the process of constructing a house has also been introduced. This theme threads throughout each chapter in the book acting as an example and demonstrating the practical application of the use of civil engineering materials. The construction of a house is probably the most practical and everyday example of civil engineering and requires all of the materials covered in the chapters in this book.

Contents

Chapter 2 Concrete 77

Chapter 1

Soils

Learning outcomes

After studying this chapter, you should be able to:

- Explain soil structure and its principles
- Discuss the importance of moisture in the soil
- Identify and perform some of the important tests relating to laboratory and field identification
- Identify soil characteristics
- Discuss compaction of soil and its importance
- Review the procedures followed when crushing rock
- Explain the purpose of stabilising soils using different methods
- Explain the importance of providing good protection to sloped surfaces
- Define the concept of reinforced earth.

1.1 Introduction

The Earth's crust consists mainly of rock and soil. Understanding the properties and characteristics of these help you to understand the behaviour of rock and soil under given conditions.

We are all familiar with the concept of a house (in whichever form it takes), so we are going to relate the various chapters in this book to the building of one. When building a house, we start with a good foundation and the knowledge that the house is built on solid ground. The house is constructed using various materials in the civil and built environment – some of which will be introduced in the various chapters of this book.

To coincide with the 'building' of the house in this book, you are required to design and draw your own house and use the examples given to relate and calculate some of the volumes required in your own home. If you have difficulty with either the design or drawings, ask your drawing lecturer for assistance. The house you design must have a maximum total floor area of $100 \, m^2$ and consist of the following:

- 3 bedrooms
- Kitchen
- Bathroom (with bath and shower facility)
- Toilet (does not necessarily have to be separate)
- Lounge
- Dining room
- Single garage (included as part of the total floor area).

Draw the house complete with floor plan, elevations and sections required to assist you in identifying relevant detail of your home.

So why is knowing all about soils so important?

Soil is not just used to support a structure, like in a foundation of a house, but it can also be used in various other applications, e.g. fill material in embankments or for reinforced earth. When installing pipelines in the ground, it is done into soil of which we need to know the properties. Road layers are constructed using certain types of 'engineered soil' whilst mixing concrete is done with a special type of sand.

Engineers and technicians rely heavily on the knowledge of these properties when designing and constructing multi-storey structures and roads. So apart from the nature of soils, we will also examine the characteristics of soil for identification purposes. We will look at soil classification, what procedures are used in soil surveys and the implications of soil compaction.

An understanding of some of the basic properties and behaviour of soil will form a basis for a more in-depth analysis in later years of study. Properties such as compaction, the construction of quarries, how to stabilise soils, protect slopes and reinforce earth will increase your insight into the very ground on which you walk.

Hands-on experience through performing specific tests will help build this understanding. These experiments should ideally be done under controlled conditions, i.e. in a laboratory, with the supervision of experienced personnel.

Activity 1

The basic idea is to mix a sample of ingredients to try and establish the 'best' in terms of consistency, strength and behaviour. Each sample will be mixed in a ratio of 1:3:3 representing one part of cement, three parts of mystery ingredient and three parts of 13 mm aggregate. The mystery ingredient must consist of the following:

- Building sand
- Cake flour
- Sugar
- Granulated plastic.

Use a cup as a measure and prepare four separate mixtures, each containing the mystery ingredient. You therefore have to prepare one sample containing one part of cement, three parts of building sand and three parts of aggregate, another with one part of cement, three parts of cake flour and three parts of aggregate, etc. Mix each of the samples separately in a large mixing bowl and observe things like the ease of mixing, the volume of each sample, the colour and consistency. Now add 1 cup measure of plain water to each mixture and mix thoroughly. Allow 24 hours for each mix to set and then test each sample for strength by crushing either by hand or using blows from a hammer. How difficult is it to break each mixture and why do you think it is so?

Please ensure that this activity is carried out in a laboratory or otherwise under controlled conditions.

1.2 Nature of soils

1.2.1 Origin

Do you still remember the composition of the earth and its rock forms?

Soil is a result of weathering, erosion and transportation of the Earth's crust over time. The uppermost layer of soil is commonly referred to as **topsoil.** Topsoil is the layer of weathered material (usually not more than 500 mm thick) containing organic matter. This topsoil sustains and supports the growth of vegetation. For engineering purposes, soil is to be considered as any loose deposit such as **gravel**, **sand**, **silt**, **clay** or a combination of these materials. Topsoil is generally removed before any engineering projects are carried out and kept at a suitable area close by (called a stockpile) so that it can be re-used.

1.2.2 Type

Various terms are used to describe particular types of soil, i.e. gravel, sand, silt and clay. Soils are divided into those with fine-grained particles and those with coarse-grained particles. Essentially, soil is divided into types according to the sizes of individual grains or particles found in the soil. Also see page 17.

Soil, in broad terms, consists of **solid particles**, **moisture** (water) and **air voids** as shown in fig 1.1.

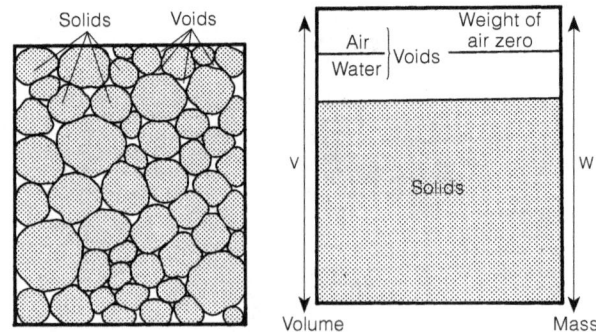

Figure 1.1 Soil, particle, moisture and air voids

Air voids are the tiny spaces between the solid particles that are filled with air. From an agricultural point of view, the more air present in the soil the better, because it allows the soil to support growth. (Plants grow better in coarse-grained soils than in fine-grained soils like clay.) From an engineering point of view, it is essential that as many as possible of the voids are removed in order to gain better strength. For this reason we use large compaction equipment to force solid particles together, thereby squeezing out as much of the air as possible, as shown in fig 1.2. This process is called **compaction**.

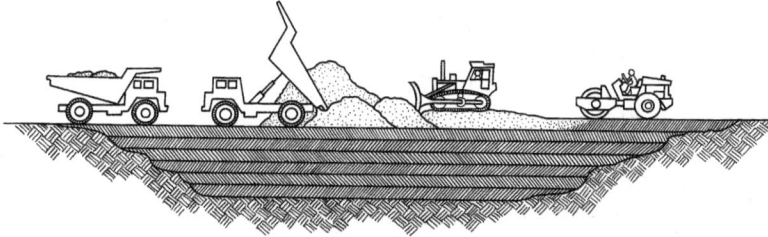

Figure 1.2 Spreading and compaction of soil

The moisture in the soil plays an important part in determining the physical properties of the soil. In a simplified way we can say that water acts as a lubricant in the soil and allows solid particles to move better over each other when compacted. Too little water will result in excessive friction, while too much water means that the solid particles will be 'swimming' in the water. To get good compaction, it is therefore important to use the correct amount of water in the soil. To determine this, certain tests are done in the laboratory. See section 1.3.3 for more on moisture content.

1.3 Identification of soil

Experienced engineers and technicians can identify a soil purely by looking at its physical characteristics. You can do the same. Look around you and you too will be able to pick up certain differences in the soil. The most obvious is that of colour – certain soils appear **black** or **grey**, some **white** (for example beach sand), while others may have a **yellowish** colour. Each of these different colours has a certain significance.

1.3.1 Field identification

Apart from colour, soil can be identified in the field using the following characteristics:
- Compressibility
- Plastic characteristics
- Dry strength
- Colour
- Smell
- Feel.

Compressibility

Take a handful of wet soil and shake it around in the palm of your hand. If there is water on your skin, the soil is defined as showing characteristics of compressibility. With the soil still in your palm, close your hand and squeeze the soil. When you open your hand, you will find that the soil remains in a lump but now appears stiff and crumbly.

An example of compressibility is when you walk on the beach close to the water and every time you place your foot on the surface, it appears as if water rushes away from your foot yet the ground does not appear to be that wet. The weight of your body is squeezing excess water out of the pores in the soil.

Plastic characteristics

To demonstrate the cohesive (ability to stick together) properties of clay, take a soft lump of clay (approximately 23 g) and roll it in your hand to form a long cylindrical shape or thread. As you continue rolling, the thread becomes stiffer because the water evaporates – this is the plastic phase of the soil. Evaporation of the water is caused by the heat generated by friction from the rolling process. Now shape the clay into a lump again and start to roll a new thread. Note how difficult it has become since the first time.

In this context, the **plastic condition** of soil describes the range of behaviour between liquid (flowing) condition and the condition at which the soil is so dry that it breaks into pieces. In the context of material behaviour, plasticity has another meaning, described on page 35.

Think about why builders have to be careful when building on clay and why concrete fences in particular become uneven and skew, while the houses they surround are free of cracks and distortions.

Figure 1.3 Distortion

Dry strength

Squeeze some soil into a lump, place it in the sun and allow it to dry for a few hours. Once dry, use your finger to apply pressure to the lump of dried soil and note the pressure necessary for it to crumble. Repeat the process for different types of soil, taking careful note of the different pressure needed for each soil to crumble. This characteristic is known as the **dry strength** of the soil. It is important that you acquire a 'feel' regarding how the types of soil differ in terms of their dry strength, so take the time to make various mixtures and compare results.

Colour

Each soil type has a distinguishing colour. For instance, dark-coloured or **black** soil indicates the presence of organics in the soil, whereas **lighter** coloured soil is inorganic. As an example, look at the colour of the soil in a pond or vlei, which appears black, indicating a soil rich in nutrients, which promotes the growth of vegetation. This colour must not be confused with pollution, which also makes the soil look black or dark in colour. On the other hand, hardly anything grows on clay, which is whitish. Iron oxide particles give gravel (ferricrete) its reddish colour. Beach sand and sand dunes are clear or white.

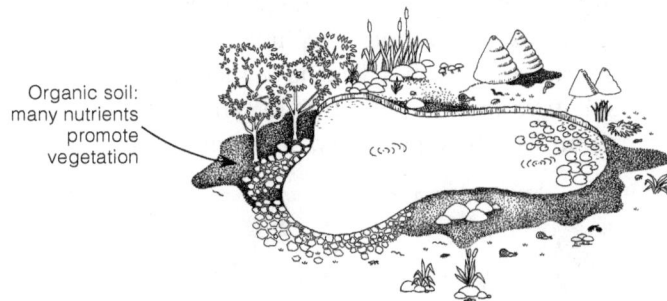

Organic soil: many nutrients promote vegetation

Figure 1.4 Pond: black organic soil provides nutrients

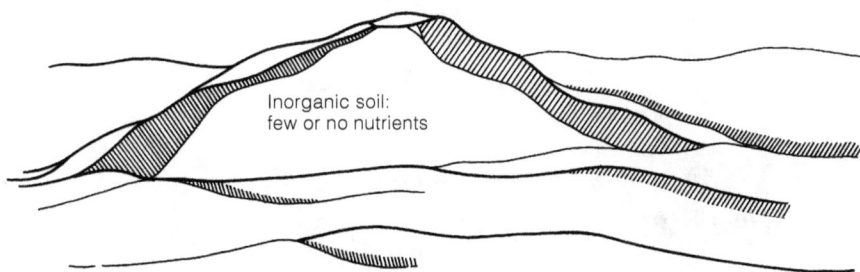

Inorganic soil: few or no nutrients

Figure 1.5 Sand dune

Smell

Certain soils exhibit a distinctive smell when disturbed, particularly organic soils. This characteristic is usually only evident when the soil is freshly disturbed, as the smell generally disappears after a while.

Feel

Touch can be used to distinguish silts from clays and from sand. Grab a handful of sand and rub it between your fingers – feel how coarse it is. See if, by using other soil samples, you can distinguish between the roughness or coarseness of each. Grab a lump of clay and you will immediately be able to tell the difference. Whereas the sand was coarse, clay feels smooth and greasy and it might also stick to your fingers. Clay also tends to dry very slowly, i.e. its water retention abilities are high. Silt on other hand dries very quickly and can be dusted off your fingers, leaving only a stain.

Activity 2

Identify an area where you will be building your house. Now get a spade and pick and **dig a square hole** approximately 1 metre wide × 1 metre long × 1 metre deep. You should be able to stand comfortably in the hole. Try to keep the sides vertical. Now examine the soil on the sides of the hole you have just dug and see how many of the characteristics that you have just read about can be applied to the soil in your own environment. The things you will be looking for are: compressibility, plasticity, dry strength, colour, smell and feel.

Figure 1.6

If you make any discoveries other than those mentioned above, write them down and compare them with observations from other people or talk to your lecturer.

1.3.2 Soil samples

In order to understand the behaviour of soil, it is necessary to study the soil in **controlled conditions** similar to when doing chemical experiments. The controlled conditions exist at various types of **civil engineering laboratory facilities** around the country. Some of these laboratories have specialised applications such as a **concrete lab**, a **bitumen lab** or a **soil lab**.

To conduct a test on soil, for example to determine its properties, a **representative soil sample** must be obtained in the field. A representative soil sample is a small amount of soil taken from the soil that needs to be tested. There are various ways of getting such samples and, depending on the method used, a disturbed sample or an undisturbed sample of soil is obtained.

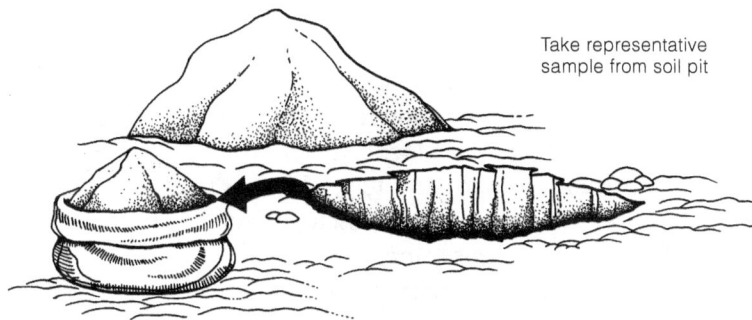

Take representative sample from soil pit

Figure 1.7 A representative sample

As the name suggests, a disturbed sample of soil is one where the soil condition has been changed. Think back to Activity 1: before digging, the soil was in an undisturbed state or condition, but as you started digging and piled the soil in a heap next to the hole, the soil then became disturbed.

It is good practice to disturb as little as possible when taking soil samples.

Undisturbed soil is usually obtained by drilling **core samples.** This method is normally used for rock.

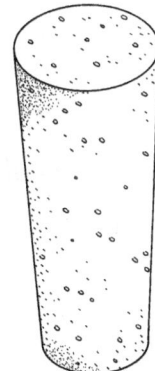

Figure 1.8 A core sample

Depending on the nature of the soil and the tests necessary, usually 40 kg is deemed to be enough soil with which to undertake testing of a disturbed sample. Once the sample gets to the laboratory, it is divided into the required volumes or mass for ease of work. Usually these masses are between 3 and 5 kg.

1.3.3 Moisture content

The procedure to find out how much water is within a soil sample is quite simple. You need a container in which to put the soil, an oven capable of maintaining temperatures between 105 and 110 °C, a weighing scale and, of course, your sample of soil. Do not allow your sample to dry out before doing this test.

Method:
Determine the mass of the empty container as well as the container with the wet soil. Write down these values. Preheat the oven to between 105 and 110 °C and place the soil and container in the oven. Leave overnight or for approximately 12 hours to dry.

You should record the following values:
- Mass of empty container
- Mass of container with wet soil
- Mass of container with dry soil.

Now determine:
- Mass of moist soil before drying
- Mass of dry soil (container with dried soil minus empty container).

Be careful when you remove the container and soil from the oven as it will be very hot.

Weigh the container with the dry soil and record the mass. To calculate the moisture content as a %, use this formula:

$$\frac{\text{Moisture}}{\text{content (\%)}} = \frac{(\text{mass of container + wet soil}) - (\text{mass of container + dry soil})}{(\text{mass of container + dry soil}) - (\text{mass of container only})} \times 100$$

or

$$\text{Moisture content (\%)} = \frac{\text{mass of moisture}}{\text{mass of dry soil}} \times 100$$

EXAMPLE 1

Determine the moisture content of the soil sample using the following information:

Mass of container = 175 g
Mass of wet soil and container = 3 125 g
Mass of dry soil and container = 2 987 g

Solution

Use the equation for the determination of moisture content:

$$\text{Moisture content} = \frac{3\,125\,\text{g} - 2\,987\,\text{g}}{2\,987\,\text{g} - 175\,\text{g}} \times 100$$

$$= \frac{138\,\text{g}}{2\,812\,\text{g}} \times 100$$

$$= 4.9\%$$

This means that for the sample of soil tested, the in situ moisture content (natural moisture in the soil at the time of testing) is 4.9%. In other words, in its natural state, the soil already has approximately 5% water in its pore structure.

1.3.4 Consistency

Consistency is that property of the soil that displays **resistance to flow**. It reflects the cohesive abilities (the ability of the soil particles to stick together) of the soil, especially that of clay materials. This property of the soil is affected by its moisture content. By changing the water content in the soil, you will find that the soil moves through various stages, i.e. from a solid phase to a liquid phase (see fig 1.9).

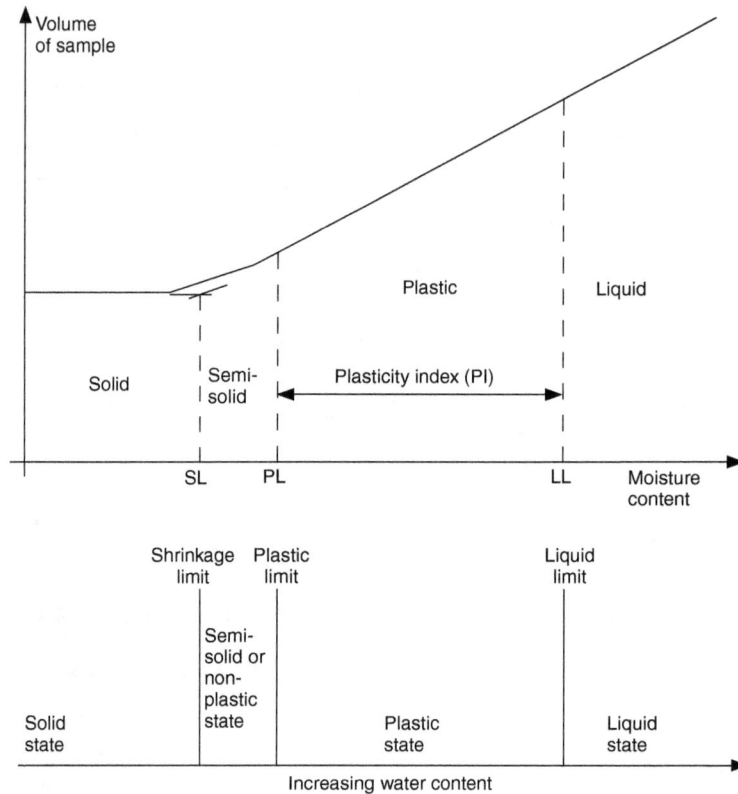

Figure 1.9 Consistency limits of soil

Consistency of the soil can be expressed using three **defined limits**. They are:

- Liquid limit (LL)
- Plastic limit (PL)
- Plasticity index (PI).

Consistency tests are used to:

- Classify soils
- Indicate soils with poor plastic qualities
- Estimate the strength of the soil (plasticity index or PI).

Which other material can also move from one phase to another?

Water changes from a liquid phase to a solid phase (when it becomes ice) and then to a vapour phase (when it is steam).

Liquid limit (LL)

The liquid limit (LL) is the moisture content, expressed as a % by weight of the oven-dry soil, at the boundary between the plastic and liquid states.

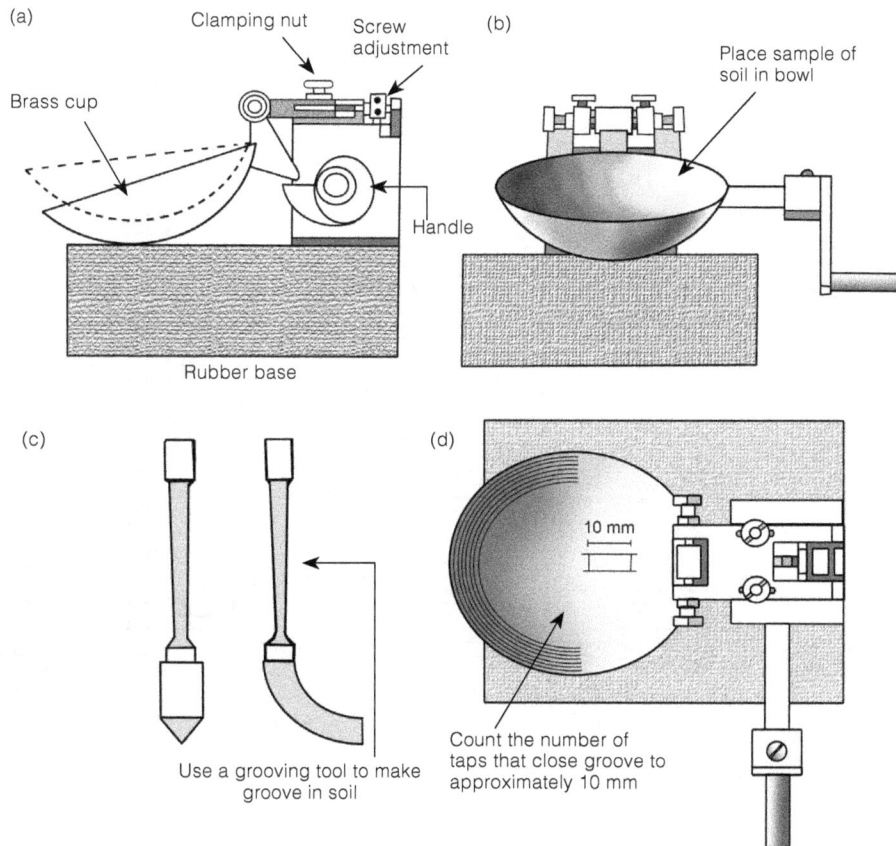

Figure 1.10 A liquid limit testing device and tools

There are three different ways to determine the liquid limit of a fine-grained soil:

- Three (3) point method
- Two (2) point method
- One (1) point method.

These points relate to the range of turns (also referred to as taps or blows) applied to the soil in the liquid limit device.

Three-point method: This method uses three points to determine the moisture content of the soil. A point is obtained from each of the following ranges:

■ 15–22 taps
■ 22–28 taps
■ 28–35 taps.

A 'best fit' line is then drawn to represent the test results.

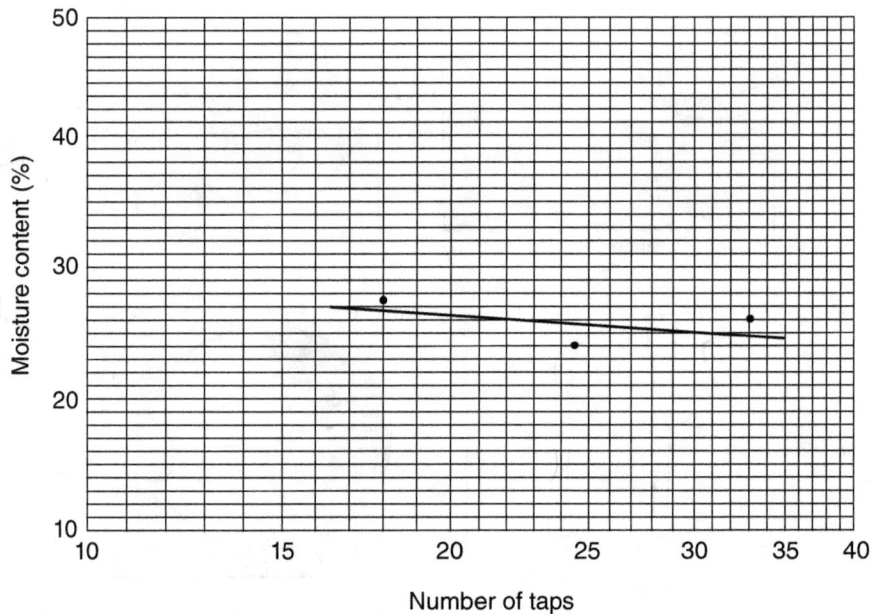

Figure 1.11 Graph showing the three-point method

Two-point method: This method uses only two consistencies:
■ A moisture content that closes the groove at 28 taps
■ A moisture content that closes the groove at 22 taps.

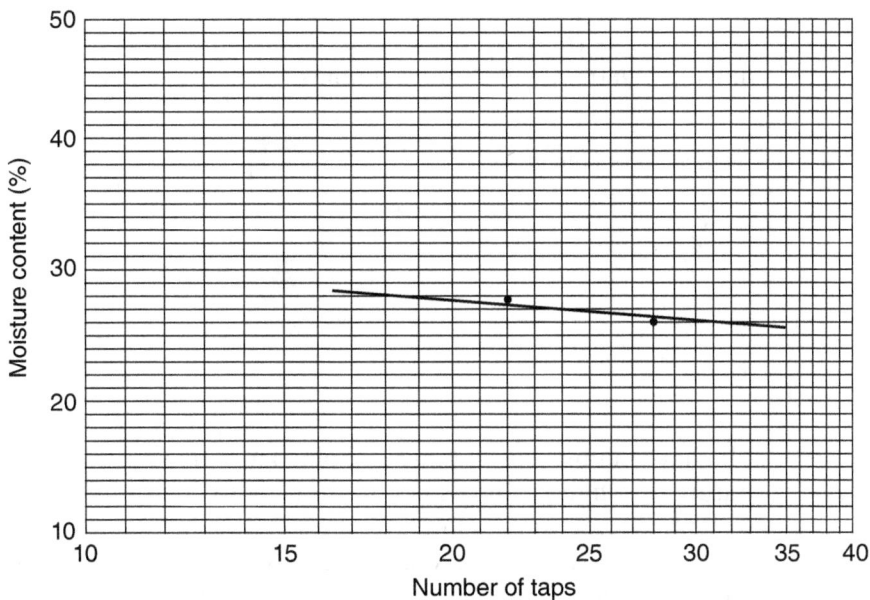

Figure 1.12 Graph showing the two-point method

One-point method: The number of taps required in this method is restricted to between 22 and 28.

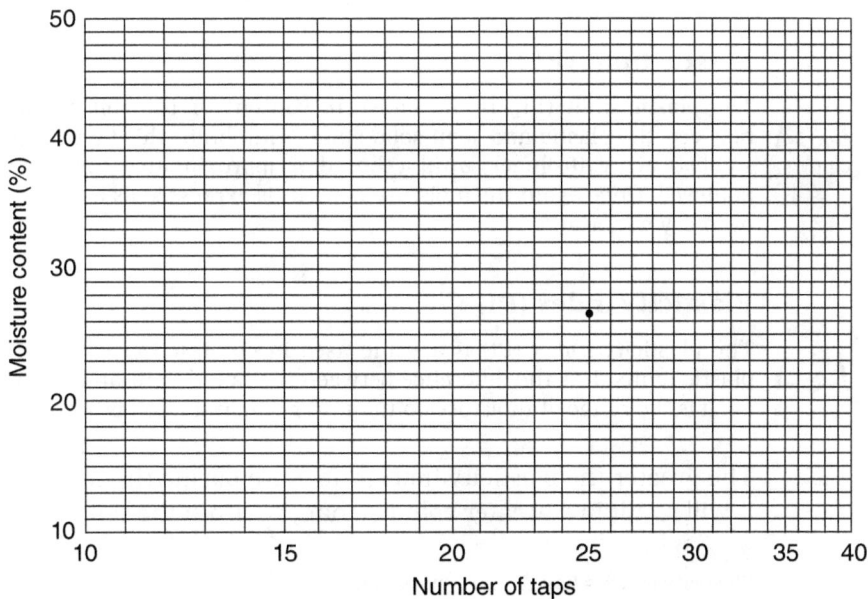

Figure 1.13 Graph showing the one-point method

Liquid limits vary, but values of 40–60% are typical for clay soils and 25–50% for silty soils. Sandy soils have no liquid limit because granular soils such as these do not exhibit cohesive properties and instead of flowing together, will either crumble or collapse when tapped.

So what does all this mean?

The liquid limit, plastic limit and plasticity index are measures of the nature of a fine-grained soil and they try to determine the amount of water needed to move a fine-grained soil through four phases, i.e. solid, semi-solid, plastic and liquid. In each phase the behaviour, consistency and engineering properties of the soil will be different. These limits are commonly referred to as the Atterberg limits and can be used to distinguish between silt and clay. Silts and clays can either shrink or expand in volume when you change the moisture content and this is an important property to take into consideration during construction because it will also alter the strength of the material.

Plastic limit (PL)

The **plastic limit** (PL) is the moisture content of the soil at the boundary between the plastic and semi-solid state. You obtain this by rolling a piece of soil between your fingers until a thread of approximately 3 mm in diameter is achieved. Once again the moisture content is expressed as a percentage of the oven dry material.

Plasticity index (PI)

The **plasticity index (PI)** is a value used in conjunction with the liquid and plastic limits. It is the difference between the liquid limit and the plastic limit, i.e. it is the range of moisture content over which the soil is in the plastic state.

The greater the plasticity index, the more plastic the soil and therefore it will result in a greater volume change within the soil.

Plasticity index = liquid limit – plastic limit
or
PI = LL – PL

1.3.5 Specific gravity test (SG)

Specific gravity (SG) is the ratio of the weight of the solid material per unit volume to the weight of an equal volume of water, under standard conditions. This property is used to calculate the **density** and **porosity** of materials. In the soil laboratory, an apparatus called a **hydrometer** is used to determine this value. This test is mostly carried out on fine aggregates.

To give you an indication of values of SG:

- Specific gravity of water is 1.0
- Most coarse aggregates used in road construction have values ranging between 2.65 and 2.70.

1.3.6 Relative density

The relative density (RD) of water equals exactly one and if a substance has a relative density of less than one, the substance will float in water. If the substance has a relative density of more than one, it will sink. An example is an ice cube, which has a RD of 0.91 and therefore floats in water.

Relative density is often also referred to as specific gravity. Specific gravity usually means relative density with respect to water.

What is a particle size distribution test?

1.3.7 Particle size distribution

A soil structure is not **homogeneous,** in other words, it is not the same throughout. You will often find soil particles of different shapes and sizes that require identification according to size. A **particle size distribution test** does exactly this – it separates the various soil particle sizes from each other.

If you take a soil with all the particles of a similar size and put them together you will find that they stack together in an orderly fashion with lots of voids (gaps) between them. If you take a soil with different fractions (sizes) and mix them all together, you will notice the voids between the soil particles are smaller. This is necessary if we want to compact the soil to obtain higher densities.

Table 1.1

clay	silt	sand	gravel	cobble	boulder

0.006   0.02   6   20   6   20

0.002 mm   0.06 mm   2 mm   60 mm   200 mm

Very coarse soils	BOULDERS		> 200 mm
	COBBLES		60–200 mm
Coarse soils	**G** GRAVEL	coarse	20–60 mm
		medium	6–20 mm
		fine	2–6 mm
	S SAND	coarse	0.6–2.0 mm
		medium	0.2–0.6 mm
		fine	0.06–0.2 mm
Fine soils	**M** SILT	coarse	0.02–0.06 mm
		medium	0.006–0.02 mm
		fine	0.002–0.006 mm
	C CLAY		< 0.002 mm

Activity 3

Take a bag of marbles and put them into a transparent container like a glass jar. Take a second bag of marbles, but this time mix it up with other marbles of various sizes and some coarse river sand before placing in a similar transparent container. Look at the voids in each container. See how much smaller the spaces between each of the particles are in the second container and how it appears to be more compact, or dense, than the first.

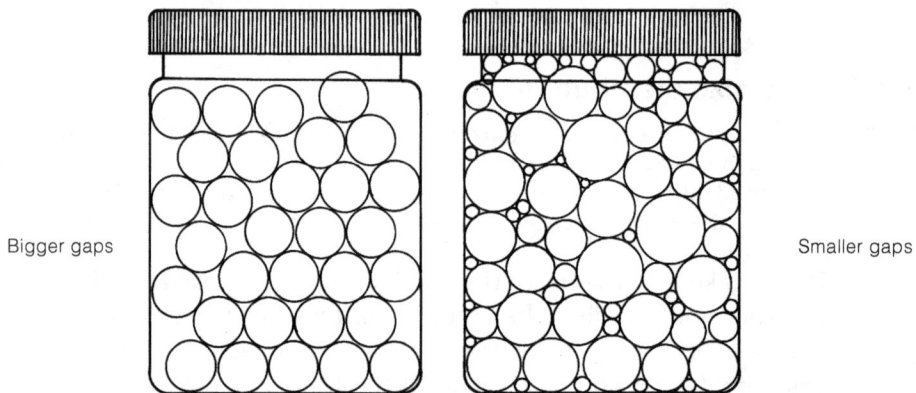

Bigger gaps   Smaller gaps

Figure 1.14 Particle size distribution

The container with the different-sized marbles and sand shows a more tightly knit structure. This is exactly what we require from soil that is going to be used in a supportive function. This soil has the ability to be compacted to a fairly high density.

In its **natural state**, soil will sometimes be of uniform size and shape, for example windblown sand. Crushed stone or rocks have irregular shapes and sizes and civil engineers need to stipulate the particular size of stone to be used. The blasting and separation processes of aggregates are dealt with in section 1.7.2.

Particle size distribution test or sieve analysis

Experiments with the various materials are done in a lab using the **particle size distribution** test. To do this test, you will require the following different-sized sieves: 63.0 mm, 53.0 mm, 37.5 mm, 26.5 mm, 19.0 mm, 13.2 mm, 4.75 mm, 2.0 mm, 0.425 mm and 0.075 mm and a pan to catch those very fine soil particles that pass through the 0.075 mm sieve.

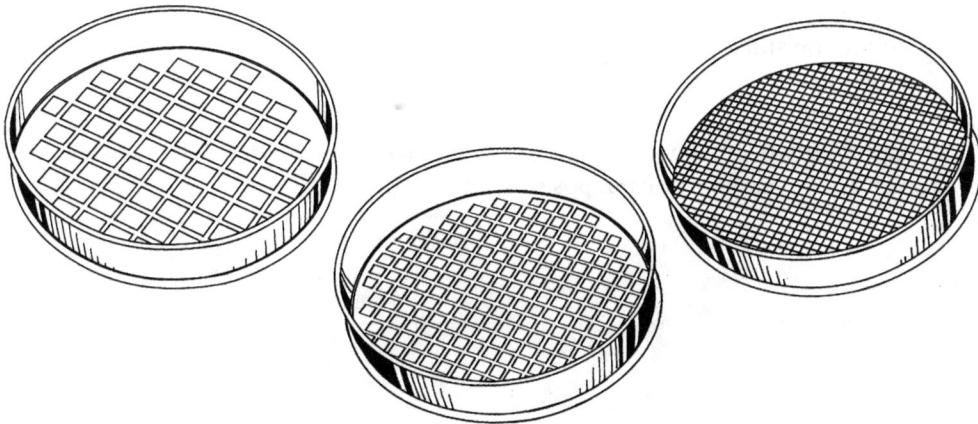

Figure 1.15 A range of sieves

The sizes refer to the openings in the sieves. In the case of a 63.0 mm sieve, the openings will be 63 mm × 63 mm. Any material larger than 63 mm, i.e. not passing through the largest sieve, is classified as rock (cobbles and boulders), and material smaller than 0.6 mm is **fine aggregate** (material less than 0.002 mm is called clay particles). See table 1.1.

The sieve analysis is carried out in accordance with test method Al as described in the TMH1 (technical methods for highways) document.

The relevant sieves are stacked one on top of the other to retain the various fractions of soil. The material retained in each sieve is then weighed and recorded on a laboratory sheet next to the relevant sieve size. The % retained is then calculated for each sieve, the values converted to % passing and these values are plotted on a graph. Usually the % values are on a vertical (y-axis) scale and the sieve sizes are on the horizontal (x-axis) scale. The horizontal scale is a log scale.

EXAMPLE 2

An oven-dried sample of gravel weighing 3 240 g is put through a sieve analysis test. The sieve sizes required for the test are as follows: 37.5 mm, 26.5 mm, 19.0 mm, 13.2 mm, 4.75 mm, 2.0 mm, 0.425 mm and 0.075 mm. A pan is also required to retain soil particles finer than 0.075 mm.

Step 1: Determine the amount of soil retained

- Stack the sieves on top of each other, ranging from the 37.5 mm sieve at the top to the 0.075 mm at the bottom.
- Place a pan below the 0.075 mm sieve to catch any material passing through the last sieve.
- Pour the soil into the top sieve, taking care not to spill any.
- Shake the stack to facilitate the movement of the soil through the various sieves.

In more advanced laboratories, you may use a special mechanical sieve shaker. However, for your purposes, shake by hand.

Do not shake too vigorously otherwise you might lose some material.

Figure 1.16 Shaking sieves

Once this is complete, take the sieves apart and weigh each fraction of soil to the nearest 0.1 grams. The results for this example are given in table 1.2.

Table 1.2

Sieve size (mm)	Mass of soil retained (g)
37.5	0
26.5	1 032
19.0	793
13.2	558
4.75	323
2.0	215
0.425	144
0.075	119
< 0.075	56
TOTAL	3 240

Step 2: Calculate the % retained in each sieve

Use the formula:

$$\frac{\text{mass of soil retained on sieve}}{\text{total mass of sample}} \times 100$$

Using the data given in table 1.2:

The mass of soil retained on the 26.5 mm sieve   = 1 032 g

The total mass of the sample   = 3 240 g

Therefore % retained is $\frac{1\,032}{3\,240} \times 100$ = 31.9%

The mass of soil retained on the 19.0 mm sieve   = 793 g

The % retained is $\frac{793}{3\,240} \times 100$ = 24.5%

Some of the calculations of % retained are given in table 1.3. Calculate the missing values to complete the table.

Table 1.3

Sieve size (mm)	Mass of soil retained (g)	% soil retained
37.5	0	0
26.5	1032	31.9
19.0	793	24.5
13.2	558	
4.15	323	
2.0	215	
0.425	144	
0.075	119	
< 0.075	56	1.7
TOTAL	3240	100.0

Check your answers with those provided in table 1.4.

Step 3: Determine % soil passing through each sieve
This calculation is fairly easy because all that is required is to use the last column on the table above and starting at the top, deduct the cumulative value from 100. The value 100 in this case is once again the % value. The % of soil passing through each sieve for this example is given in table 1.4. Calculate the missing values to complete the table.

Remember to check your answers in each step.

Table 1.4

Sieve size (mm)	Mass of soil retained (g)	% soil retained	% soil passing
37.5	0	0	100
26.5	1032	31.9	68.1 (100 − 31.9)
19.0	793	24.5	? (68.1 − 24.5)
13.2	558	17.2	
4.75	323	10.0	
2.0	215	6.6	
0.425	144	4.4	
0.075	119	3.7	1.7 (5.4 − 3.7)
< 0.075	56	1.7	0 (1.7 − 1.7)
TOTAL	3240	100.0	

Step 4: Plot the values on a graph

Use a sheet of semi-logarithmic (semi-log) graph paper (see fig 1.17), and the first and last columns in table 1.4. Mark the sieve sizes on the horizontal scale (x-axis) and the % passing on the vertical scale (y-axis). Plot the values and connect all the points to produce a particle size distribution curve. Fig 1.18 shows that different soils exhibit different grading curves.

Remember that the x-axis is a semi-log scale and the y-axis is a normal scale.

- Soil A is a **well-graded** material. Notice the smooth curve of the line, taking in all of the sieve sizes.
- Soil B is a **gap-graded** material. See the resemblance of an s-shape. The more pronounced the shape, the higher the number of missing sieve sizes. The material is of such a nature that some of it is not retained in one or more of the sieves.
- Soil C is commonly known as a **uniform** material, where the material is mostly retained in only one or two sieves, e.g. beach sand, building sand, etc. Notice the almost vertical shape of the curve.
- Particles smaller than 0.002 mm are tested by means of a flotation test.

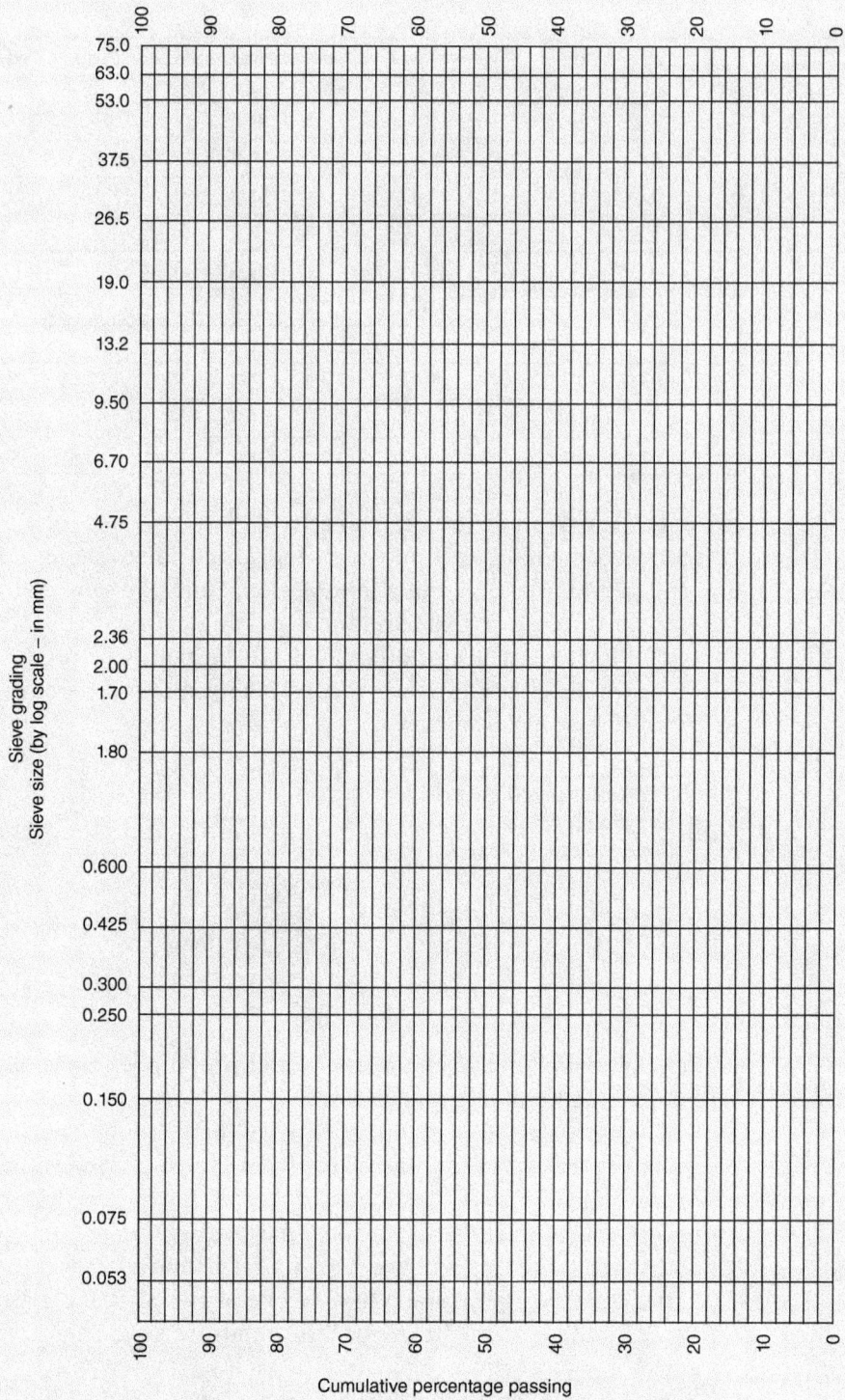

Figure 1.17 Semi-logarithmic graph paper used for sieve analysis

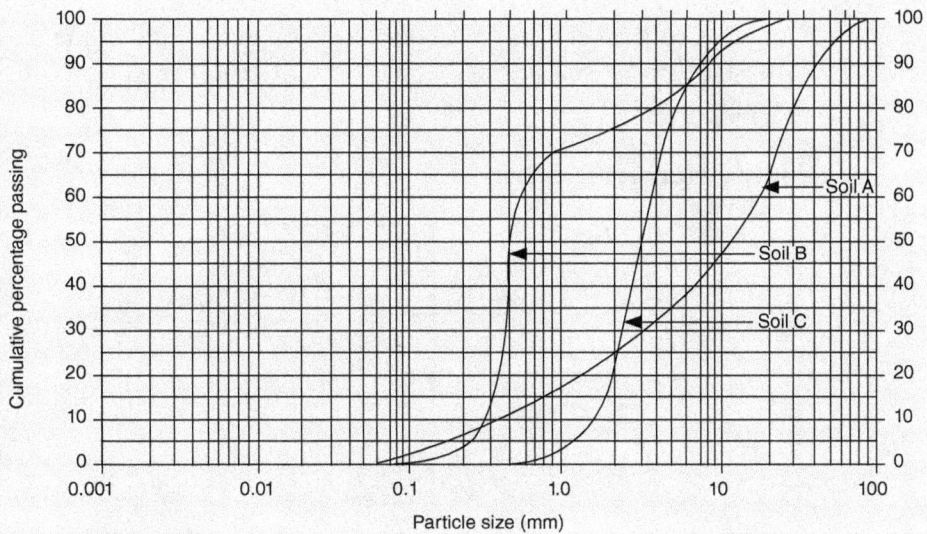

Figure 1.18 Soil grading curves

Depending on the type of material being tested or the required specification, you may get variations to these three mentioned. The engineer or technician will determine what the best material is for the given situation.

What characteristics of soil are most suitable to build on?

Bearing capacity

In geotechnical terms, bearing capacity is the ability of soil to support the loads applied to the ground. Strip footings (as is common to house construction) are generally classified as shallow foundations – these footings transfer the load of the entire house structure onto the soil. It is therefore important that the soil be of a good supportive nature so that the house does not collapse due to foundation failure.

A soil derives its strength from two sources, namely cohesion and frictional resistance. The cohesion is largely a function of the fine-grained material and its ability to 'cement or bond' particles together. Frictional resistance is derived from the larger particles of a soil 'rubbing or interlocking' against each other and this contact results in friction. Factors that would influence the strength of soil are soil composition, structure of the soil, condition of the soil and loading.

Grading characteristics

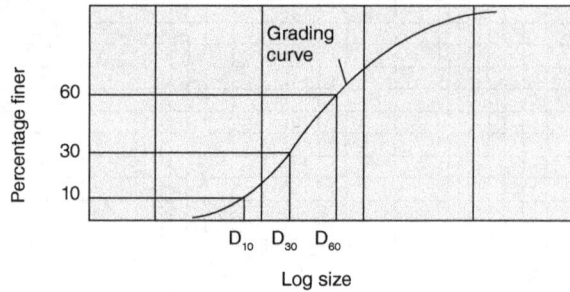

A grading curve is a useful aid to soil description. Grading curves are often included in ground investigation reports. Results of grading tests can be tabulated using geometric properties of the grading curve. These properties are called grading characteristics.

First of all, three points are located on the grading curve:
D_{10} = the maximum size of the smallest 10% of the sample
D_{30} = the maximum size of the smallest 30% of the sample
D_{60} = the maximum size of the smallest 60% of the sample

The **coefficient of uniformity**, C_u is a crude shape parameter and is calculated using the following equation:

$$C_u = \frac{D_{60}}{D_{10}}$$

where D_{60} is the grain diameter at 60% passing, and D_{10} is the grain diameter at 10% passing.

The **coefficient of curvature**, C_c is a shape parameter and is calculated using the following equation:

$$C_c = \frac{(D_{30})^2}{D_{10} \times D_{60}}$$

where D_{60} is the grain diameter at 60% passing, D_{30} is the grain diameter at 30% passing, and D_{10} is the grain diameter at 10% passing.

Once the coefficient of uniformity and the coefficient of curvature have been calculated, they must be compared to published gradation criteria. Both C_u and C_c will be 1 for a single-sized soil.

$C_u > 5$ indicates a well-graded soil.
$C_u < 3$ indicates a uniform soil.
C_c between 0.5 and 2.0 indicates a well-graded soil.
$C_c < 0.1$ indicates a possible gap-graded soil.

More examples of soil classifications are:

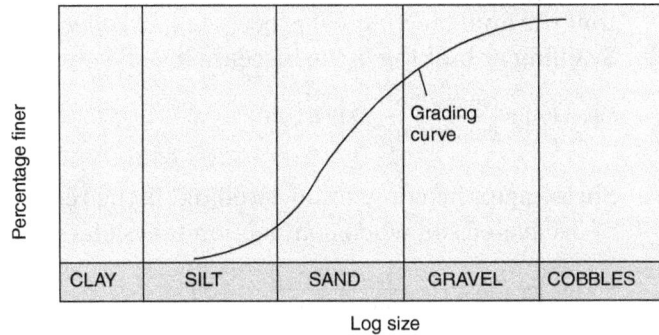

Typical grading curves:

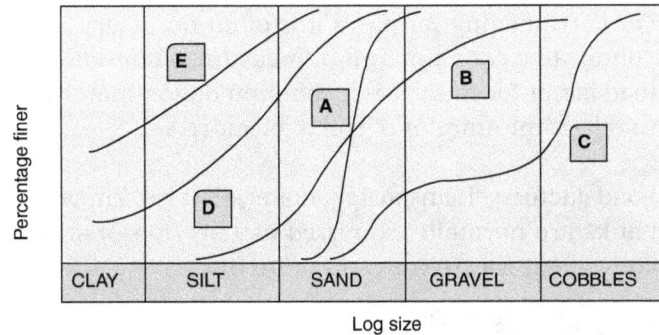

Both the position and the shape of the grading curve for a soil can aid in its identity and description. Some typical grading curves are shown in the figure:

A – a poorly graded medium SAND

B – a well-graded GRAVEL-SAND (i.e. equal amounts of gravel and sand)

C – a gap-graded COBBLE-SAND

D – a sandy SILT

E – a typical silty CLAY

1.3.8 Soil volume characteristics

There are three different stages or principle conditions to describe the volume of a soil, namely bank, loose or compacted.

■ **Bank condition** (Bm^3) refers to the soil in its natural state, i.e. an

undisturbed sample of soil will be referred to as bank volume
$1\,\mathrm{Bm^3} = 1\ \mathrm{m^3}$

- **Loose condition** ($\mathrm{Lm^3}$) is loose material after excavation, e.g. a disturbed sample of soil, which can be in transit (moving from one location to another). Swelling of the soil takes place in this stage
- **Compacted condition** ($\mathrm{Cm^3}$) is the reduced volume after transporting, placing and compaction has taken place. Compaction here means that the final mechanical process (e.g. a roller) has been applied
- **Swelling or bulking** is the increase in soil volume from bank to loose

$$\text{Swell (\%)} = \left[\frac{\text{mass/Bm}^3}{\text{mass/Lm}^3} - 1\right] \times 100$$

- **Shrinkage**, the converse of swelling, is the reduction in the volume of excavated soil when compaction takes place.

$$\text{Shrinkage (\%)} = \left[1 - \frac{\text{mass/Bm}^3}{\text{mass/Cm}^3}\right] \times 100$$

Conversion of material volume

For earthmoving purposes it is often necessary to convert all material volume to a common unit of measure, often $\mathrm{Bm^3}$, using in this case the load factor formula. It is a common denominator used as an indication to reflect the amount of soil to be moved.

Load factors: Transported material (also known as haul) and spoil banks are normally expressed as $\mathrm{Lm^3}$ (loose) and the load factor is a convenient way to convert $\mathrm{Lm^3}$ to $\mathrm{Bm^3}$.

$$\text{Load factor} = \frac{\text{mass/Cm}^3}{\text{mass/Bm}^3}$$

or

$$\text{Load factor} = \frac{1}{\left(1 + \dfrac{\text{swell}}{100}\right)}$$

Shrinkage factor: This factor is used to convert $\mathrm{Bm^3}$ to $\mathrm{Cm^3}$ (compacted).

$$\text{Shrinkage factor} = \frac{\text{mass/Cm}^3}{\text{mass/Bm}^3}$$

or

$$\text{Shrinkage factor} = \left(1 - \frac{\text{shrinkage}}{100}\right)$$

EXAMPLE 3

A soil mass is 1 163 kg/Lm³, 1 661 kg/Bm³ and 2 077 kg/Cm³.

a. Calculate the load and shrinkage factor
b. How many Bm³ and Cm³ are contained in 593 300 Lm³ of this soil?

Solution

a. Load factor = $\dfrac{Lm^3}{Bm^3}$

$= \dfrac{1\ 163}{1\ 661}$

$= 0.70$

b. Shrinkage factor = $\dfrac{Bm^3}{Cm^3}$

$= \dfrac{1\ 661}{2\ 077}$

$= 0.80$

For 593 300 Lm³
Bank volume = 593 300 × 0.70 = 415 310 Bm³
Compacted volume = 415 310 × 0.80 = 332 248 Cm³

Remember that we are dealing with **volumes** of soil, **not areas**.

A **spoil bank** is where the pile of material is long in relation to its width, i.e. in relation to B.

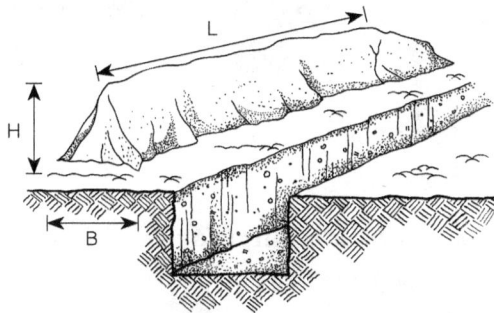

Figure 1.19 A spoil bank

Spoil is a term used to describe excavated material that can be kept for later reuse. A **spoil pile** is where the material is dumped from a fixed position to create a conical shape.

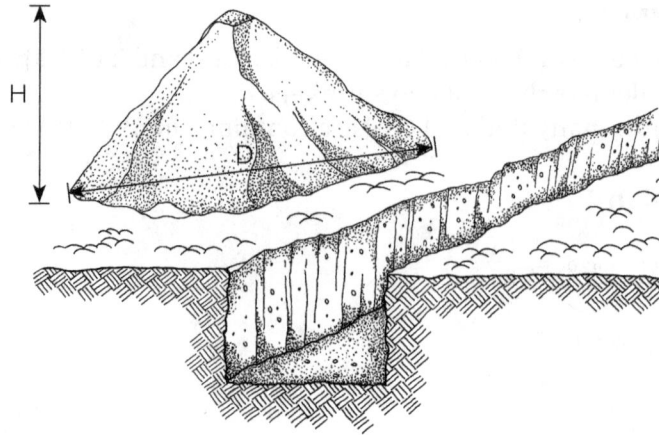

Figure 1.20 A spoil pile or stock pile

To calculate the dimensions of spoil banks or piles, it is necessary to convert from Bm³ to Lm³. We can then calculate bank or pile dimensions given the angle of repose of the soil. The **angle of repose** is the steepest angle a soil can be banked, without slipping or collapsing. Typical values for angle of repose are given in table 1.5.

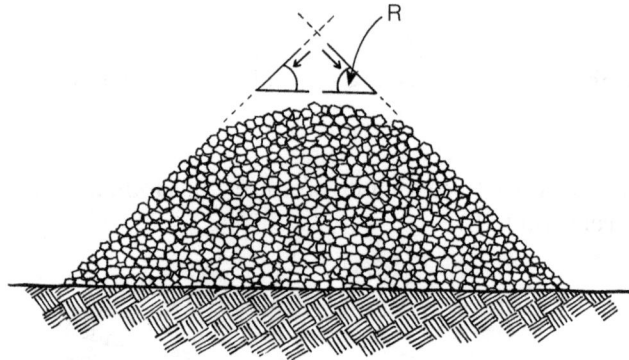

Figure 1.21 Angles of repose of material

Dimensions of a triangular spoil bank

Volume = section area × length

$$V = \frac{BH}{2} \times L$$

$$B = \left[\frac{4V}{L\tan R}\right]^{\frac{1}{2}}$$

$$H = \frac{B\tan R}{2} \quad \text{(see fig 1.19)}$$

where B = width of pile
       H = height of pile
       L = length of pile
       R = angle of repose
       V = loose volume of soil (Lm³)

Dimensions of a conical spoil pile

Volume = $\dfrac{1}{3} \times$ (area of base × height)

$$V = \dfrac{1}{3} \times \pi R^2 \times H$$

$$D = \left[\dfrac{7.64V}{\tan R}\right]^{\frac{1}{3}}$$

$$H = \dfrac{D \tan R}{2} \quad \text{(see fig 1.20)}$$

Table 1.5 Typical angles of repose

Material	Angle of repose
Clay	35^0
Common earth – dry	32^0
Common earth – moist	37°

EXAMPLE 4

Find the base width and height of a triangular spoil bank containing 76.5 Bm³ if the length is 9.14 m, the angle of repose is 37⁰ and the swell is 25%.

Solution

There is a 25% increase due to swell
$\therefore (1 + \frac{25}{100}) = 1.25$

To convert Bm³ to Lm³:

76.5 Bm³ × 1.25 = 95.6 Lm³

$$\text{Base width (B)} = \left[\dfrac{4V}{(L \tan R)}\right]^{\frac{1}{2}}$$

$$\text{Base width (B)} = \left[\dfrac{4 \times 95.6}{9.14 \times \tan 37^\circ}\right]^{\frac{1}{2}}$$

$$= 7.45 \text{ m}$$

$$\text{Height (H)} = \dfrac{(B \tan R)}{2}$$

$$= \dfrac{(7.45 \times \tan 37^\circ)}{2}$$

$$= 2.80 \text{ m}$$

EXAMPLE 5

Remember the house introduced earlier? In order to build the foundations, we need to excavate the trenches for all the areas where foundations are needed. Typically this is underneath all the walls of the structure. For the house in the example, the total length of wall which needs excavation amounts to 350 m. Assuming that the trench is 1 200 mm wide and 900 mm deep, what will be the total volume of loose material (Lm^3) if a bulking factor of 1.1 is applied to the excavated material? Also calculate the width and height of the triangular spoil bank if the angle of repose is 35°.

Solution

Hint 1: Convert all the measurements to a single unit.

Hint 2: The place where you will be excavating is undisturbed, i.e. natural state.

Hint 3: If 1 equals 100% of a fully compacted material in its natural state, what would 1.1 relate to in terms of swell of that same material?

Step 1: Convert all measurements to the same unit, e.g. metres

1 200 mm/1 000 mm	= 1.2 m
900 mm/1 000 mm	= 0.9 m

Step 2: Calculate bank volume:

Bank volume	= 350 m × 1.2 m × 0.9 m
	= 378 m³

Step 3: Calculate loose volume:

Loose volume	= 378 m³ × 1.1 (bulking factor)
	= 415.8 m³

Step 4: Calculate spoil bank dimensions:

Base width (B)	= $[(4 \times 415.8)/(350 \text{ x tan } 35°)]^{\frac{1}{2}}$
	= 2.605 m
Height (H)	= $(2.605 \times \text{tan } 35°)^2$
	= 0.912 m

Therefore, when excavating for the foundations to the structure, you will have 415.8 m³ of soil from its natural state. The question is, what happens to all of this material? The answer will be dealt with later in this chapter.

Self-evaluation 1.1
1. Complete the sentences:
   a. The uppermost layer of soil is commonly referred to as _____.
   b. Soil can be divided into solid particles, _____ and _____.
   c. Clay has _____ properties.
   d. Consistency is that property of soil that displays _____.
   e. In the liquid limit determination, a value corresponding to _____ taps is read off the graph.
   f. _____ is defined as the moisture content at the boundary between the plastic and semi-solid state.
   g. There are three different principle conditions that can describe the volume of a soil, i.e. _____, _____ or _____.
2. State whether the following statements are **true** or **false**:
   a. One grain of soil is equal to an equal number of particles.
   b. One can identify a soil by merely looking at the physical characteristics of it.
   c. Compressibility refers to dry soil that crumbles under the slightest pressure.
   d. Usually soil samples of up to 5 kg are taken as being a representative sample from the field.
   e. Liquid limits are expressed as %.
   f. When doing the plastic limit test, approximately 40 g of soil is rolled in your hand.
   g. The particle size distribution test separates the various soil particle sizes from each other.
3. Complete the following calculations:
   a. Find the base diameter and height of a conical spoil pile that contains 89.2 Bm^3 if the angle of repose is 32° and the swell of the soil is 12%.
   b. Complete the sieve analysis for the data in table 1.6.
      ■ Calculate the % passing for this soil.
      ■ Plot the particle size distribution curve (using semi-logarithmic graph paper).
      ■ From the graph, give your comments about the characteristics of this soil.

Table 1.6

Sieve aperture (mm)	Mass retained (g)
63.0	35.3
53.0	185.6
37.5	287.1
26.5	172.6
19.0	104.4
13.2	94.8
4.75	98.0
2.0	73.2
0.425	115.7
< 0.425	183.9
TOTAL	

1.4 Classification of soil

Soils throughout the world have different characteristics but, in order for engineers and scientists to identify these soils, different institutions have devised different systems, each based on a different behaviour. In devising these systems, soil investigators have tended to group soils together on the basis of some common characteristics.

Definitions

A **class** refers to a group of soils resembling one another. The purpose of **classification** is to place a soil sample in a group or class that can be internationally understood.

Soil classification systems are based on particle sizes found in the soil mass. A classification system is used:
- As a means of identifying various soil groupings
- To provide a consistent, internationally recognised common language for soil investigators to interchange information about similar soils
- As a basis for decisions on further tests required for solving a particular engineering problem.

Most systems recognise three main soil types or classifications (grain refers to the individual mineral particles in the soil):
- Coarse-grained/cohesionless soil, e.g. sand
- Fine-grained/cohesive soil, e.g. clay
- Organic soil, e.g. peat.

Coarse-grained soil: The sieve analysis test is usually carried out on these soils. This is because the size of the particles and the proportions of the different sizes have an important effect on the behaviour of the soil. Coarse-grained material does not necessarily mean large particles;

umgeni sand or beach sand can also be considered a coarse-grained material.

Fine-grained soil: The properties of fine-grained soil are affected by the water content. Cast your mind back to an earlier section where it was stated that clay is considered a fine-grained material. The criteria for testing are those of **consistency** and **plasticity**. Consistency is the tendency of the particles to stick together and plasticity is the ability to deform without rupture or breaking (also see description of plasticity on page 6).

Two of the most common soil classification systems used in South Africa are:

- Casagrande classification system
- Unified soil classification system. You will study these classifications in Geotechnical Engineering II.

1.5 Soil survey procedure

A soil survey forms an important part of the preliminary engineering survey and provides the engineer with crucial information with regard to the soil and ground water conditions. Soil tests could be surface testing or sub-surface tests like drilling boreholes.

Can you think of two reasons why soil and ground water conditions affect projects in crucial ways?

1.5.1 Personnel

It is important that qualified personnel skilled in the various testing methods and techniques are used to perform surface and sub-surface tests. The importance of doing the tests properly cannot be over-emphasised. Consistent and accurate test results are crucial as designers of civil engineering projects rely upon the accuracy of these results in their final design of the project.

Figure 1.22 A laboratory technician and assistant

1.5.2 Equipment

Various types of testing equipment are used, depending on the tests that need to be performed. For tests that are done in the laboratory, TMH1 identifies the apparatus necessary for each. In the field where drilling needs to be done, we distinguish between a **hand auger** or a **mechanically operated** one. Hand augers are up to 150 mm in diameter and are usually used up to a depth of 3 m, depending on the soil conditions. Mechanical borers are used for much deeper boreholes.

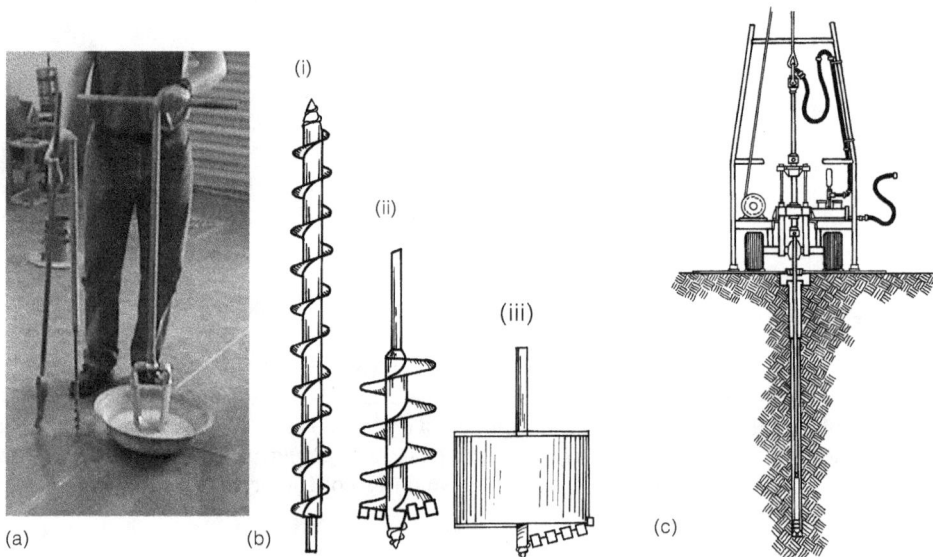

Figure 1.23 (a) Using a hand auger (b) (i) Continuous-flight auger
(ii) Short-flight auger   (iii) Bucket auger (c) A mechanical auger

With the advances in technology, more use is made of other techniques of soil investigations such as **satellite**, **sonar** and other **geophysical** methods.

1.6 Compaction

In civil engineering, soil is commonly used as a construction material, both as a **supporting** medium or in its own right. In the latter case, soil is often transported from one location to another to be used in construction, usually as a fill material in the construction of **embankments**. Such transported soil has to be treated in order to bring it from a loose disturbed state into a more densely packed or compressed material. The process used to achieve this is known as compaction.

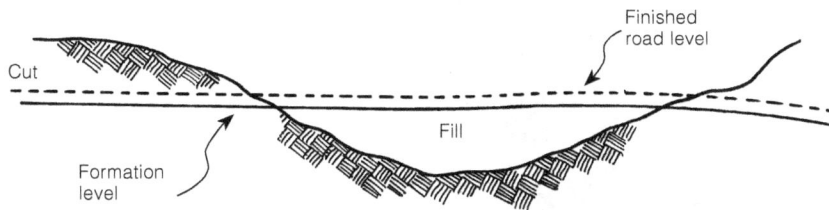

Figure 1.24 Soil profile showing cut and fill

In compaction, the soil particles are packed closer by forcing out the air and water, usually by using mechanical construction plants such as **rollers**, **tampers** and **vibrators**. Compaction results in a **reduction in the volume** of the soil due to the expulsion of air from the voids.

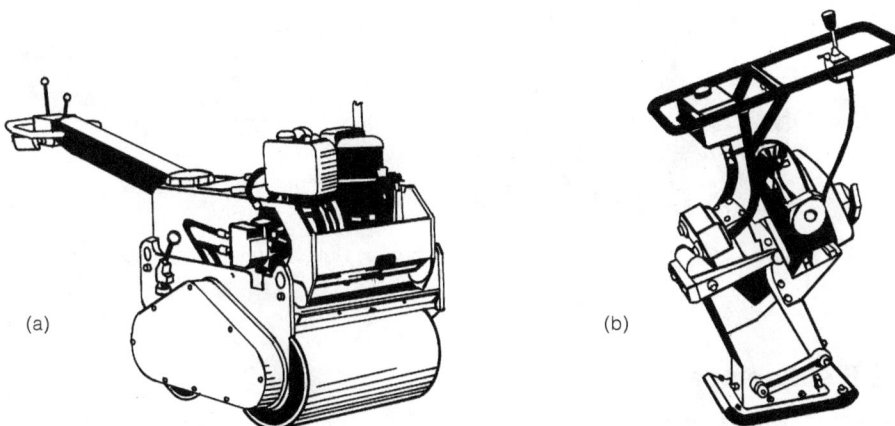

Figure 1.25 (a) Roller compactor (b) Plate compactor (tamper)

1.6.1 Measurement of compaction

As compaction creates a denser soil mass, the effect of compaction can be measured in terms of **dry density**, i.e. the mass of solid material per unit volume of the soil. **Bulk density** and the moisture content of the soil are used to calculate dry density. As compaction is concerned with air void content, it can also be used to calculate the volume of air.

The effect of compaction can be measured both in the field as well as the laboratory. Laboratory tests determine the degree of compaction achievable when using a standard amount of compactive effort, with a view to understanding the probable degree of compaction obtainable in the field. The most common tests used to determine the strength and density of the soil are the MOD AASHTO, Proctor and CBR (California bearing ratio) tests. You will learn more about these tests when doing Geotechnical Engineering II.

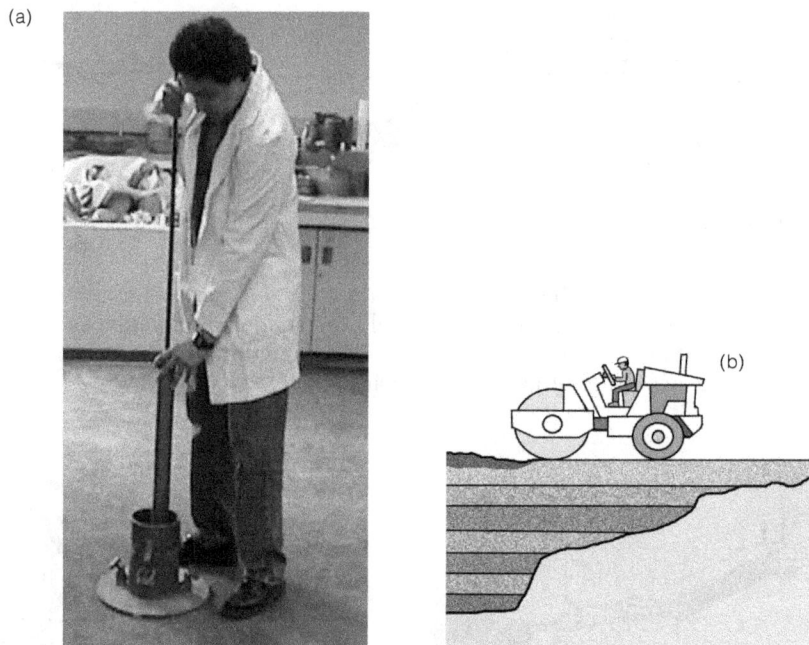

(a)

(b)

Figure 1.26 (a) Compacting in a laboratory (b) Roller compacting a soil layer

1.6.2 Field compaction

A variety of equipment can be used for the compaction of soil in the field, ranging from hand tampers to heavy vibrating rollers. Normally the soil is compacted in layers, the thickness of which depends on the soil type and compactive effort.

(a)   (b)

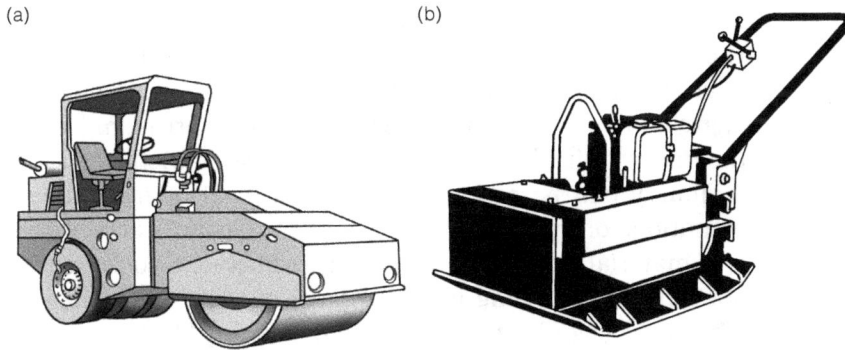

Figure 1.27 (a) A heavy roller (b) A plate compactor

What are the two methods of ensuring compaction to a required specification?

There are two methods of ensuring compaction to a required specification:

■ The first stipulates the **compaction procedure** to be followed, specifying the type of compaction plant to be used as well as the number of passes for each layer – similar to a recipe when making food or baking a cake. The detailed method is clearly given and it is hoped that the final compaction is the required value.

■ The second method indicates the **end result required**, usually in terms of the 'in-place' dry density of the compacted soil, i.e. via the Proctor, MOD AASHTO or CBR tests. How this compaction is arrived at is left entirely to the contractor, as long as the final result is achieved

In some cases a combination of both methods is specified.

1.7 Quarries

Quarries are specific areas that have been identified as a source of a particular material, for example coarse aggregate (stone and sand) and fine aggregate (clay). Remember that the general classification for aggregates has two main categories, coarse and fine, but within these general classifications you will find sub-classifications. An example of this is clay particles which are considered very fine particles, but are still categorised under the general category of fine aggregates, whereas sand is coarser (when compared to clay) and technically considered as the finer fraction of a coarse aggregate. The same will apply to the coarse aggregate classification where large boulders are technically considered the upper limit of the coarse fraction classification (refer to table 1.1). A quarry can be compared to a gold, coal or iron mine, as similar operations of extraction (in this case rock) are carried out. To obtain any material, explosives are used to blast the rock into manageable portions. When living near a quarry, you will often hear loud bangs when explosive charges are set off. Before any blasting takes place, strict safety procedures are followed – but more about this later. Most sand, silt and clay comes from soil pits. Sand dunes may be a source for building sand. Soil pits are generally not subjected to the same blasting and extraction processes as quarries, as the material used is more accessible through the use of normal excavation techniques. Bulldozers and front-end loaders are the most common earthmoving equipment used in soil pits. There are strict safety procedures to be followed before any blasting takes place and dealing with or handling of explosives is best left to the experts.

1.7.1 Development of a quarry

It is important that soil studies are carried out to identify the type of rock as well as the uses thereof, before actual mining takes place. Proper layout planning of the site is also important to determine:

- Quality control
- Geological influences
- Slope angles
- Excavation of the materials
- Traffic on haul roads
- Quarry planning in relation to the layout of the quarry.

The **assumptions** made in the early stages of planning will require **validation** during the initial stages of quarrying. The benefits of proper geo-technical and production planning will ensure optimisation during the life of the quarry.

A geographical contour map (a drawing that shows the levels of the ground above sea level, as well as the type of material beneath the surface) is used to make the quarry as accessible as possible and to locate buildings and roads on the site. Once the area has been decided upon, a tacheometric survey is carried out and a contour map is drawn showing all rock outcrops.

Figure 1.28 Conducting a tacheometric survey

The amount of rock that can be extracted from the quarry is determined by geologists. This is the **deciding factor** of the actual size of the quarry as well as its usefulness. Test holes are drilled into the ground on a gridline basis to determine where and to what extent solid rock will be found. The prevailing **wind direction** also plays an important part in the location of the site. The wind should not blow dust into a residential area.

What can be done to minimise dust from a quarry?

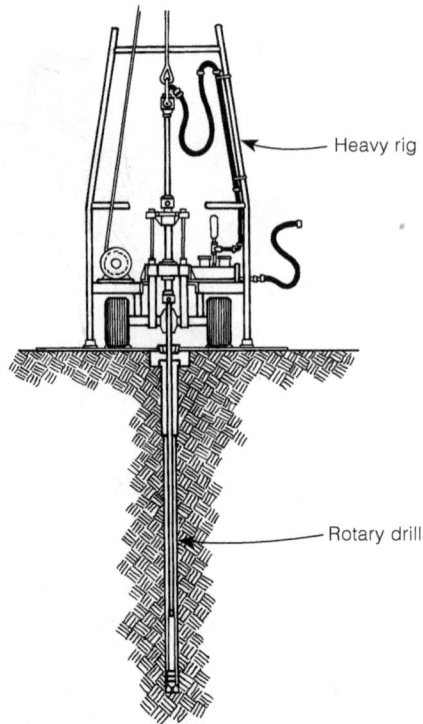

Figure 1.29 Test hole drilling

Roads

In order to gain access to and from the workface as well as to other areas of the site, it is vital that proper haul roads are constructed. These roads should:

- Be well maintained and designed, as otherwise they can cause bottlenecks and increase costs
- Slope down from the workface to the crushing plant. The reason of course is that when trucks come from the workface, they are fully loaded with raw material and it is easier and cheaper to go downhill than uphill
- Have an even, well-drained surface and be wide enough to enable vehicles to pass each other easily to prevent hold-ups
- Not have any sharp curves (corners) as this could be dangerous
- Have a level loading area to avoid unnecessary problems
- Not have a gradient of more than 1:9 as anything steeper will make it difficult for fully laden trucks to transport the material (to visualise a gradient of 1:9 use your ruler and mark out 9 cm along the horizontal. At this mark, measure 1 cm vertically. Now connect these points to form a triangle. On a bigger scale, use metres in

exactly the same manner. This sloping line represents a gradient of 1:9)
- Be kept relatively smooth, i.e free from rocks and potholes, to prevent damage to the tyres of the vehicles used.

Figure 1.30 Tyre punctures: a result of an uneven road surface

Dust control

Dust is always a major problem on a site, particularly if it is located near residential areas. The Department of Mining has, since 1986, made it a requirement that all quarries are regularly checked and sampled as a means of dust control. If there is no control over the amount of dust, it could have serious health and safety implications for the people employed there. In most cases, a water truck moves around the site spraying water to control the amount of dust.

Rehabilitation requirements

It is now a requirement of the law that all mining excavations must plan an **extensive rehabilitation programme** before operations start. **Rehabilitation** is the process whereby reconstruction of the area takes place so that it can blend in with the surrounding landscape once the quarry is exhausted (used up).

Definitions

Why do you think rehabilitation programmes are a requirement of the law?

Note the following regulation contained within the Mines and Works Act (5.1.4): 'No sand shall be extracted from the bank of any stream, river, dam, pan or lake, except with the written permission from the Inspector of Mines and upon such conditions as the said Inspector may prescribe.'

1.7.2 Blasting

Blasting is the process whereby the original rock deposit is fragmented, using explosive devices or drilling methods, so that it can be transported from the rock face to the crushing plant.

Definitions

Once a suitable area has been identified as being rich in a particular type of rock, drilling takes place. The intervals at which these holes are drilled, as well as the depth of each hole, are determined by the amount of rock to be blasted and the capacity of the explosives. These calculations are done by the blasting experts.

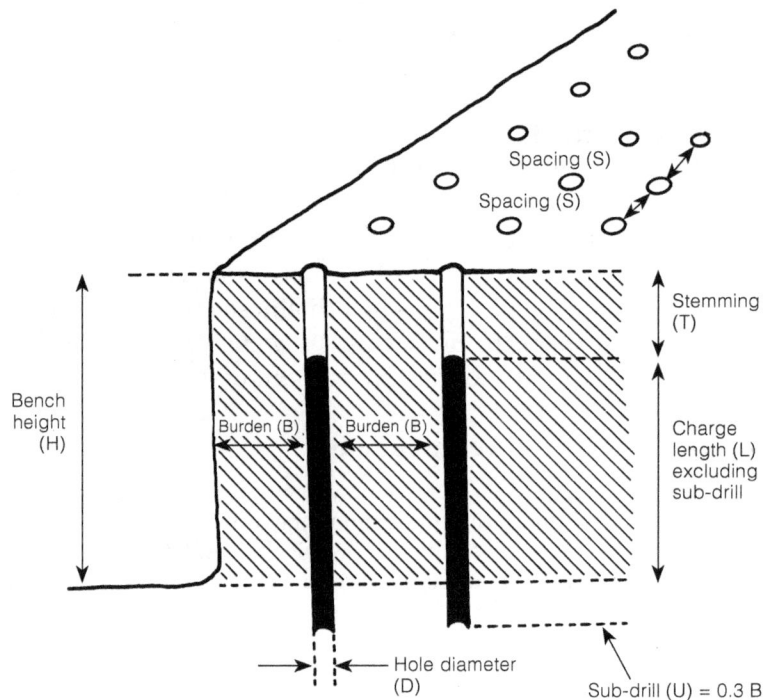

Figure 1.31 Blasting terminology

Stemming

Stemming is the process of covering, closing or plugging the hole in readiness for blasting – just like placing seeds in the ground and covering them with dirt.

Definitions

After the explosive charge is placed in the hole, it is primed and then stemmed to the **collar** (the rim of the hole) with sifted sand or clay. This is done to prevent the explosive charge from jumping out of the hole once it has been set off. Good stemming also helps to reduce **flyrock** (rock fragments that become airborne during the blasting process) and improves the results of the blasting.

The optimum stemming length can only be determined through experience and is influenced by:

- Rock conditions
- Explosive strength
- Charge length
- Bench height
- Hole diameter
- Explosive density
- Flyrock control.

Explosives

An explosive is a mixture of chemicals which, when subjected to shock, reacts extremely rapidly to release energy and gasses at very high pressures. Explosives usually consist of oxidisers, i.e. oxygen-excessive materials (fuel) and oxygen-deficient materials (sensitisers).

Two types of energies are released:

- Shock energy
- Heave energy.

Types of explosives

Several types of explosives are in use, some of which are listed here. Each of these has a different method of arriving at its charge density:

- **Pourable/pumpable explosives:** This type of explosive is like a gel or pliable piece of clay and completely fills the cross-section of the hole. Typical examples are ANFO Bulk Powergel and Energan
- **Rigid cartridges:** This group includes ammon gelignite and ammon dynamite. These can be used under water or in areas exposed to prolonged periods under water
- **Soft-packaged explosives:** These may consist of aluminium particles plus an inert compound that is suspended in water, which allows it to be used in wet conditions. It is very expensive. An example is Powergel Cold Emulsion
- **Cordtex detonating fuse:** This is essentially a thin continuous column of highly explosive materials contained in a flexible sheath. It detonates at high velocity when initiated and in turn causes the detonation of any explosives with which it makes contact. It is used in the charging of deep boreholes, i.e. deeper than 3 m

■ **Detonating relays:** These are used to time the blast by delaying charges in the rows of holes.

Blast design

Blast design considerations are:
■ Powder factor (the mass of explosives required to break one cubic metre of in situ rock)
■ Fragmentation required
■ Rock type
■ Explosive type
■ Bench height
■ Hole size
■ Stemming length.

The powder factor required to break a given rock type depends on:
■ The strength of the rock
■ Geological preconditioning, i.e. joints and faults
■ The strength and characteristics of the explosive
■ The required fragmentation.

Blasting patterns

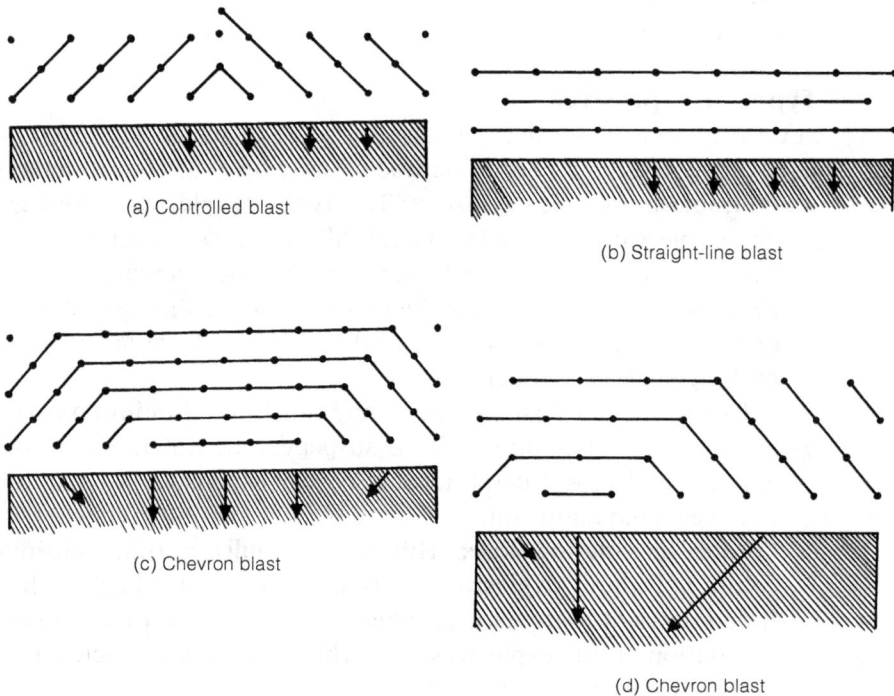

(a) Controlled blast

(b) Straight-line blast

(c) Chevron blast

(d) Chevron blast

Figure 1.32 Blasting patterns

The pattern used when drilling holes plays a big part in the fragmentation as well as the loading operation afterwards. Straight-line blasting with staggered holes give a well-spread blast. This is normally used when the loading machinery has a limited reach, for example a front-end loader.

A chevron pattern tends to pile rock into a heap. This increases fragmentation to a certain degree but can make loading more difficult. However, when only limited space on the quarry floor is available, this method is ideal.

Generally, in practice, a combination of the two patterns is used. This prevents too much piling up of rock and the chevroned sides bring out the sides which stops back-breaking and fracture on the remaining face. The position and nature of the blast can be fairly well controlled by combining the straight-line and chevron patterns, adding the vectors (the vectors giving direction and size of each section) as indicated in fig 1.32.

Figure 1.33 A front-end loader at the work-face

Figure 1.34 A front-end loader loading a truck

Hazards of blasting

> *What blasting hazards can you think of?*

- **Flyrock:** Stone fragments that are airborne directly after blasting can cause damage and serious injury. It is therefore important that all the necessary safety precautions are adhered to as well as moving personnel as far from the blast zone as possible. Causes of flyrock are poor drilling alignment, out-of-sequence firing, using too much explosive or loose rocks
- **Airblast:** Complaints about blasting are often wrongly directed at ground vibrations, when airblast is the real culprit. The level of airblast is largely determined by changes in weather conditions as well as by varying degrees of confinement of explosive charges. Because of the noise levels evident from the blast, the Bureau of Mines has recommended a maximum noise level of 134 decibels. This value is lowered to 126 decibels in a residential area. The two main causes of excessive airblast are detonation of unconfined charges and blown-out shots resulting from inadequate stemming and poor confinement
- **Ground vibrations:** Whenever you are exposed to an explosion, you will experience two physical sensations, namely vibration and noise. Vibrations are often slight and can be discarded as being nothing but could still cause alarm owing to their unexpected nature. On the other hand, pavement breakers and heavy traffic cause significant vibration.

Blasting vibrations are caused by the transmission of shock waves from the explosive charge through the materials being blasted. Have you ever dropped a stone in a pool of water and seen the mushroom effects as the circles widen? Something similar happens after every blast.

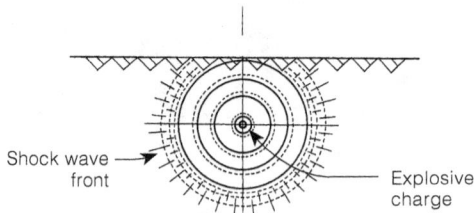

Figure 1.35 Blasting pressure waves move outwards like the ripples in water

Genuine complaints arising from blasting vibrations are very rare. Cracked plaster on a wall is usually the basis of complaints, but this is a common experience in all residential structures and can result from causes other than blasting. There are many reasons why buildings crack, other than through vibrations, for example they are built on clay or fill material, the foundations are not wide enough or compacted enough, or there is not enough cement in the concrete.

It is important to inform people when blasting will take place, so that they can take the necessary precautionary measures.

Rules in blasting

A warning that blasting is about to take place must be sounded and is required by law, almost like when standing at a level crossing or a train station where the train siren announces the arrival of the train. It also acts as a warning so that people are not caught unawares.

It is better to have a few large blasts than many smaller ones. The larger blasts should however be kept within the required vibration limit.

It is very important to minimise the noise levels. Noise levels cause windows and furniture to rattle, giving the incorrect impression of dangerous vibration levels.

Measurements of ground vibrations must be recorded and kept as proof that the required limits were not exceeded.

Secondary blasting

Although one of the general aims in primary blasting is good fragmentation of the rock, further breaking is necessary to enable oversize rocks to be handled by the loading equipment and the primary crusher. The methods used are given in table 1.7.

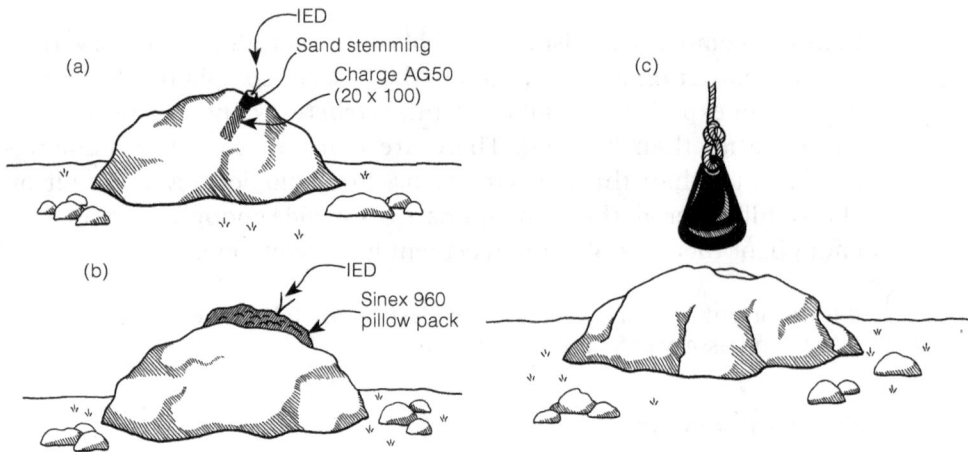

Figure 1.36 The three different methods of secondary blasting (a) Pop-shooting (b) Lay-on charge (c) Drop ball

Table 1.7 Methods of secondary blasting

	Popping	**Mud blasting**	**Drop ball**
Method	A small diameter hole is drilled about one third of the way into the boulder, charged with explosives and fired by detonator	The charge is placed on top of the boulder (in a depression, if possible) and covered with mud or clay	A cast-steel weight is dropped from a crane onto the oversize stone
Advantages	■ A small quantity of explosive is required ■ No undue scattering of rock occurs ■ A large number of explosives can be fired at once	■ No drilling required ■ Very little flyrock ■ The job is quickly done	■ No drilling required ■ No explosive costs ■ No clearing of equipment and personnel from blast zone before blasting ■ No bulldozing of scattered rock ■ No interruptions to the loading operation
Disadvantages	■ Flyrock ■ Rock scatter on the floor of the quarry ■ Labour and compressor required to drill the holes	■ Requires four times as much explosives as used in popping ■ Noise nuisance ■ The possibility of charges being set off or dislodged by prior explosions	None worth mentioning

1.7.3 Loading and transport

The loading and transport equipment required at the quarry is proportional to the amount of rock being quarried. For loading, front-end loaders and power shovels are normally used. Rock is generally transported by dump-trucks or dump-trailers pulled by tractors. In some instances, conveyor belts may also be used.

The loading and haulage procedure must run smoothly at all times, i.e. good access and haulage roads must be available and you should prevent crowding of the quarry platform.

1.7.4 Handling of explosives

Safety is of prime importance when quarrying and the Explosives Act must be adhered to at all times.

Under no circumstances must **finished (used) explosives** be stored in the building where explosives are manufactured. All finished explosives must be moved to the factory magazine or sent away from the factory immediately.

Explosives must be free of all **foreign matter**. Whenever there is danger that such matter may be present, the ingredients used to make the explosives must be examined, sifted or treated to remove them.

Misfires

A misfire is a drill hole (or part thereof) in which the blasting material or any portion thereof charged into the hole has failed to explode or the contents of which are unknown.

Misfires occur when:

- There are cutoffs in the detonating fuse. This can be due to bad charging of the drill holes
- The drill hole is much larger in diameter than the actual cartridge of explosive. This is because in a larger hole, the fuse may not be in contact with all the cartridges.

When a misfire is located after a blast, only authorised personnel may go near the site.

1.7.5 Crushers

The four types of crushers used in the crushing of stone are outlined in table 1.8.

Table 1.8

	Jaw crusher	Gyratory crusher	Impact crusher (hammer mill)	Roll crusher
Type	■ Single-toggle type ■ Double-toggle type	Inverted pestle and mortar	Hammer mill	■ Single-roll crusher ■ Two-roll crusher
Purpose	■ Secondary crusher ■ Primary crusher	Primary crusher	Primary crusher	■ Primary crusher ■ Secondary crusher
Output and process	Produces a high proportion of stone similar to the setting of the jaws	Stone is crushed between a stationary inverted bowl and a moving cone mounted on a vertical axis	Stone is broken by impact of hammers mounted on a high-speed shaft	Stone is crushed between the rotating cylinder and a fixed plate or between the two rollers in the case of a two-roll crusher
Advantages	Good cubical stone sizes and shape	Shape of stone is flakier and a finer aggregate is obtained than with the jaw crusher	■ Breaking is by impact, not pressure ■ Best stone shape and sizes obtained using this crusher	■ Single-roller is used for soft material, e.g. coal, clay, limestone ■ Two-roll crusher used for soft or hard rock
Disadvantages	Fine aggregate is not well graded and shaped			

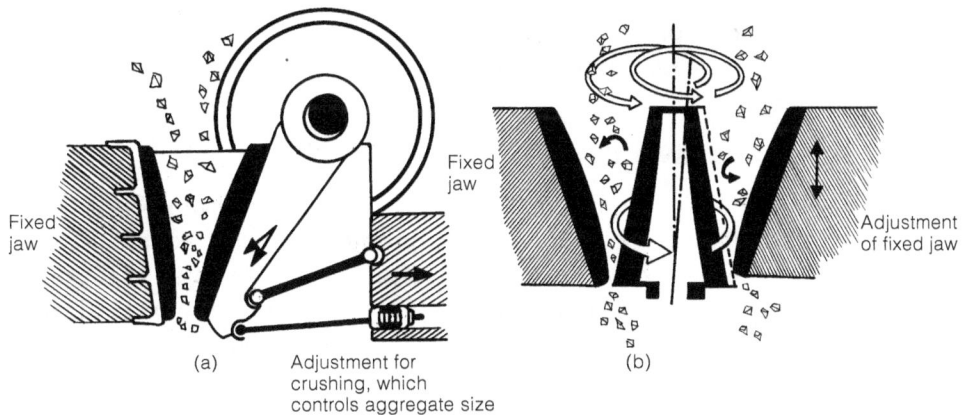

Figure 1.37 (a) A jaw crusher (b) A gyratory crusher

Selection of crushers

There are several factors that make the plant manager decide to use a particular crusher:

- **Characteristics of the material:** For thin, flaky rock, a gyratory crusher is suitable. Such rock produces low quality aggregate and should not be used. If the rock is non-abrasive or soft, a roll crusher will do. For massive material or hard, tough rock, a jaw crusher is usually best
- **Average daily/hourly capacity required:** This must normally be greater than the overall crushing plant capacity as interruptions to feed are inevitable. This reserve capacity should be from 25–50%. Normally in a quarry there is little reserve between the quarry and the crushing plant, mainly because the large boulders and stones cannot be stored
- Size of the feed
- Size of the product
- Quarry equipment
- Method of feeding the crusher
- Drilling and blasting methods.

1.7.6 Quarry products

Typical quarry products are:

- Crusher sand: < 4.75 mm
- Asphalt aggregate: 69 mm
- Crusher run/quarry waste: 4.75–19 mm
- D.I.Y. concrete aggregate: 8–11 mm
- Concrete construction aggregate: 19–26 mm
- Railway ballast: 50 mm
- Mass concrete aggregate: 63–150 mm
- Gabion basket/Reno mattress packing: 150–300 mm

- Hand stone for grouting: 200–300 mm
- Rip-rap for dams and scour protection: 450–900 mm.

1.7.7 Codes and laws

The codes used are the SANS 1083 **Aggregates from Natural Sources.** The existing document is a compromise that the technical committee responsible for it had hoped would satisfy the variation in the quality of material and differing requirements of specifying authorities throughout the country. There is also a relationship between SANS 1083 and other related documents, for example SANS 10100 and the SANS 1200 series. At the moment the quarrying industry has formalised quality systems by using the Code of Practice ISO 9000 series quality systems in conjunction with SANS 1083.

Activity 4

Ask your lecturer to arrange a site visit to a stone quarry. Record your findings in the form of a report and compare it to the notes in this book.

Self-evaluation 1.2

1. Complete the sentences:

   a   Soil classification systems are used to _____ various types of soil.

   b   A _____ refers to a group of soils resembling one another.

   c.   The process used to densify soil is known as

   _____.

   d.   _____ are specific areas that have been identified to supply or provide a particular material.

   e.   The inclines of the road must not exceed a gradient of

   _____.

   f.   _____ describes the process of covering, closing or plugging the hole before blasting.

   g.   A _____ is a drill hole or part thereof that has failed to explode.

2. State whether the following statements are **true** or **false**:

   a.   Consistency tests are done on coarse-grained material.

   b.   Anyone can perform the various soil testing methods.

   c.   Compaction is measured in terms of dry density.

   d.   The loading area at the blast area must be kept reasonably level.

e.   Explosives are mixtures of chemicals.
f.   Fragmentation is influenced by the type of drilling machine.
g.   It is better to have a few small blasts than one or two larger ones.
3.   Answer the following:
a.   Name the two methods of ensuring compaction to a required specification in a soil layer.
b.   Discuss five issues that need to be taken into consideration when providing a road in a quarry.
c.   Why is it necessary to rehabilitate an area that has been quarried?
d.   Name the two types of energy released when blasting.
e.   How do we experience a blast?
f.   Why is it important not to enter an area that has just recently been blasted?
g.   Name the four types of secondary crushers that can be used in a quarry.

1.8 Stabilisation

To quote the chairman of the American Highway Research Board, 'a stabilized fill, subgrade, basecourse or road surface is one that will stay put, and stabilizing is the process by which it has been made that way'.

Some of the important advantages of stabilisation are:
■ The strength of the material is increased
■ Durability and resistance to the effects of water are improved
■ Wet soils can be dried out
■ The workability of clayey materials can be improved
■ Cemented layers possess good load-spreading properties.

Do you know what stabilisation is?

Soil stabilisation can be considered to be any process that may improve the condition of a soil and make it more stable. Often the construction of road bases and surface courses with already stable materials is not considered part of soil stabilisation, but any treatment used to improve the strength of a soil by reducing its susceptibility to the influence of water and traffic is. This is true whether the process is performed in situ or applied to the soil before or after it is placed in the road or embankment.

Soil stabilisation can be divided into four main groups:

■ Mechanical stabilisation
■ Cement stabilisation
■ Lime stabilisation
■ Bitumin stabilisation.

1.8.1 Mechanical stabilisation

Mechanical stabilisation is the most widely used method that relies on the inherent properties of the soil material for stability. If a soil cannot be made stable simply by compaction, then additional soil or other aggregate materials may be added to produce a mixture with the required stability characteristics. Other methods included as part of mechanical stabilisation are thermal procedures involving freezing and heating of the soil.

When a granular structure such as a road base has the property of resistance to lateral displacement under load, it is said to be **mechanically stable**. In mechanically stabilised soils, this resistance is provided by the natural forces of cohesion and internal friction within the soil. **Cohesion** is mainly associated with the **silt** and **clay** content of the material while **internal friction** is a characteristic of the **coarser particles**.

Types of mechanical stabilisation

Mechanical stabilisation can be achieved by treatments such as compaction, consolidation, and electrical and thermal methods, or with the aid of additives such as soil aggregate, chlorides and lignin. The latter two stabilisation techniques (chlorides and lignin) are not generally used in South Africa.

Stabilisation by compaction

An adequate state of compaction of the soil in pavements, subgrades or embankments is essential. Increasing the state of compaction of a soil stabilises it by increasing its strength, reducing the possibility of settlement (collapse) and minimising changes in moisture content.

The compaction which can be obtained in the field during the course of construction is dependent on the following:

■ Moisture content at which compaction is carried out
■ Soil type
■ Compactive effort
■ Means by which compaction is carried out.

The selection of the proper compaction equipment and method is vital. Essentially there are four ways by which soil can be compacted:

- Using **heavy weights** to press the particles together, e.g. smooth-wheeled rollers
- **Kneading** of the soil while at the same time applying pressure, e.g. sheepsfoot roller. The pneumatic-tyred roller has a compaction action that is a cross between that obtained with a smoothwheel roller and a sheepsfoot roller
- **Vibrating** the soil so that the particles are shaken together into a compact mass, e.g. vibrating rollers
- **Pounding** the soil so that the particles are forced to move closer together, e.g. plate compactors, rammers.

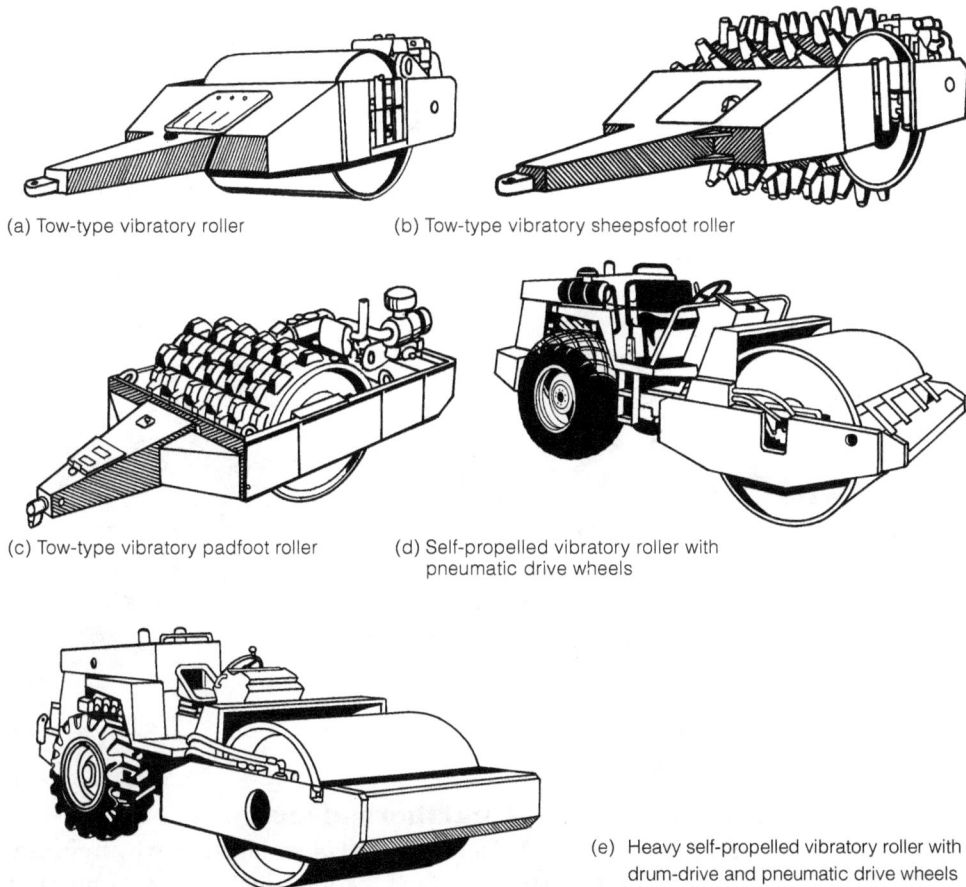

(a) Tow-type vibratory roller

(b) Tow-type vibratory sheepsfoot roller

(c) Tow-type vibratory padfoot roller

(d) Self-propelled vibratory roller with pneumatic drive wheels

(e) Heavy self-propelled vibratory roller with drum-drive and pneumatic drive wheels

Figure 1.38 Different types of rollers

Stabilisation by consolidation

Consolidation is the **pre-loading** of a soil, usually clay or peat, prior to the construction of an earth embankment, with time allowed for the soil to be partially or completely consolidated under the increased load.

To demonstrate the effect of pre-loading, take an orange and squeeze it gently. When you take your finger away, you can see a slight indentation or deformation at that point. Should you leave the orange for a while, it will resume its natural shape. By applying added pressure to the orange, to the point where the juice inside the orange oozes out, you will find that the orange will no longer be able to resume its round shape.

Similarly, soils are loaded by placing an embankment. Gradually the moisture in the soil will be squeezed out and settlement/compaction will take place. The embankment will eventually be removed and might, for example, be replaced by a structure. This process is called pre-loading of the soil.

Pre-loading by direct construction of an embankment is by far the most common way of pre-stressing a poor soil. On large construction sites it is usually preceded by a thorough soil investigation to determine the height of the fill as well as the length of time it should be left in place. In the course of construction, vertical side drains may be constructed to accelerate the rate of consolidation.

Figure 1.39 Pre-loading an embankment

Stabilisation by electrical and thermal methods

A thermal method of soil stabilisation is that of solidification by **freezing,** as was the case in the tunnel construction during the Huguenot tunneling project. This process relies on the fact that freezing the water in the pore spaces of a soil gives a very high strength to the material or makes it more workable.

Soil heating, on the other hand, relies on the fact that when a clay soil is heated it loses water. This water loss is due to the removal of absorbed and interlayer moisture held by the clay minerals above certain temperatures, resulting in changes to the actual structure of the clay minerals. The changes in structure brought about by heating clay soils result in a hard, durable road material. It has a relatively low quality, but can be useful where high-quality materials are not available. Heating causes the plasticity index to decrease while permeability is increased due to shrinkage of the clay by desiccation and the formation of cracks and fissures.

Soil-aggregate stabilisation

Soil-aggregate stabilisation is the altering of the gradation, i.e. changing the proportions and sizes of sand and stone particles of a soil. The correct proportioning of coarse aggregate to fine is essential in obtaining the desired interlocking and subsequent compaction. There are two methods used (refer to section 1.3.7):

- Blending materials by gradation
- Blending materials by plasticity index.

1.8.2 Cement stabilisation

The following are used as stabilising agents:

- Cement
  - Ordinary cement (OPC)
  - Sulphate-resisting cement (SRC)
  - Blast-furnace cement (PBFC)
  - Cement 15 SL PC 15SL
  - Cement 15 FA PC 15FA
  - Blends of ground granulated blast-furnace slag (GGBS) and OPC
  - Blends of OPC and fly ash
  - Masonry cement
- Lime
  - Slaked lime
  - Unslaked lime
- Blends of milled granulated blast-furnace slag and lime
- Blends of fly ash and lime.

It is essential that precautions are taken to protect workers from the harmful effects of lime. Unslaked lime can cause severe burns when it comes into contact with moist skin and can be dangerous if it comes into contact with the eyes and mucous membranes. Slaked lime is not quite as dangerous.

What safety precautions must be taken with lime?

The following safety precautions should be taken:
- Wear suitable protective clothing
- Apply petroleum jelly (vaseline) to exposed parts
- Wear safety glasses or goggles
- Wear respirators if there is a possibility of inhalation
- Shower or bath after work
- Wash off immediately if there is any contact with the skin.

Soil suitable for treatment with cement

Cement is effective in stabilising medium- to low-plasticity materials but it is difficult to treat fine, clayey materials owing to the high cement content required and the difficulty of pulverising the soil and mixing in the cement. Lime is a better stabilising agent for fine, clayey material. Fine, single-sized windblown or dune sands can be stabilised with cement, but large amounts of cement are usually required.

Compaction characteristics

The addition of a stabilising agent to a soil usually results in an increase in moisture content and a decrease in density. The increase in moisture is caused by the flocculating effect and by the water demand of the stabiliser. The reduction in density is caused by the development of early bond strength between particles that form loosely bonded aggregations. These aggregations may not be broken down by compaction and the longer the compaction is delayed after the cement comes into contact with water, the greater the strength of the loose aggregations and the greater the reduction in density and strength.

Factors affecting strength

The bond strength between the particles is influenced mainly by:
- The type of soil
- The amount of cement
- The density of the compacted material
- Compaction moisture content
- Curing conditions.

Cracking in cement-treated layers

Cracks in cement-treated layers cannot be avoided and must be accepted as an essential feature of cement treatment. Cracking may cause structural and maintenance problems and therefore must be controlled so as not to have an adverse effect on the performance of the pavement layer.

There are essentially two types of cracks in cement-treated layers:

- **Initial cracking** caused by drying shrinkage or thermal effects or both, but not by traffic
- **Traffic-associated cracking** caused by traffic over-stressing the cement-treated layer. It can occur anywhere in the pavement layer.

1.8.3 Lime stabilisation

Two basic chemical reactions take place when lime is mixed with a soil:

- A fairly rapid and sometimes instantaneous amelioration that may involve the exchange of ions
- A pozzolanic reaction that takes place over a period of time ranging from a few minutes to several months and longer.

Soils suitable for treatment with lime

Lime is most effective when there is a sufficient amount of clay in the soil to react with the lime. A rule frequently used is that when the PI > 10, apply lime in order to reduce the PI and increase the strength. If the PI <10 use cement.

Do you still remember what PI stands for?

Natural gravels, such as certain sandstones, calcretes, decomposed granite and decomposed dolerites, are examples of materials whose strength may increase considerably when treated with lime, in spite of PIs that may be below 10. Materials such as calcretes react strongly with lime and may have higher strengths than when treated with cement. Most plastic materials can be treated with lime, although some clay minerals react more strongly than others.

Some features of lime stabilisation

- The density of lime is about half that of cement and therefore a larger volume of lime is easier to mix than an equal mass of cement
- The reaction that takes place when lime makes contact with the damp soil usually has the effect of reducing the plasticity and making the soil more friable and more workable
- The cementing or pozzolanic reaction of lime can be a relatively slow process and in such cases there is not that same urgency to complete the construction as in the case of cement
- It is easier to use lime in wet or damp conditions

Cracking in lime-treated layers

Initial cracking develops in lime-treated layers, with the cracks forming rectangles like in cement-treated layers. However, the cracks are generally narrower and less extensive.

Laboratory test methods

California bearing ratio (CBR) – TMH1-A7: The CBR gives an indication of shear strength of the soil. The CBR is not a sensitive test when high-strength materials are tested, nor is it suitable for testing cemented materials.

Unconfined Compressive Strength test (UCS) – TMH1 A14: This test is most commonly used for evaluation of cemented materials. In this test, soil samples are subjected to increasing loads until failure occurs. Therefore, the UCS of a stabilised material is the load in kiloPascals (kPa) required to crush a cylindrical specimen 127.0 mm high and 152.4 mm diameter to total failure at a rate of application of load of 140 kPa/s. It has been shown that a relationship between CBR and UCS does exist but it depends on soil type and quantity, and type of stabilising agent used.

Wet/dry test – TMH1 A19: This test is used to determine the durability of a cemented material, especially when subjected to severe wetting and drying and/or severe traffic loading.

Tensile test: The direct tensile test, the indirect tensile test and the flexural test are three tests that can be used to determine the tensile strength of cemented materials.

Choice of strength test

The **CBR** is a useful test for estimating the strength of modified materials. The UCS test is easy to perform and it has been used extensively for mix design and quality control. The tensile strength represents the loading conditions in the field. The flexural test is particularly suitable for structural pavement design.

1.8.4 Bitumen stabilisation

Bitumen emulsion (see chapter 3) has successfully been used in the past as a stabilising agent when working with problem soils. Cationic (positively charged) and anionic (negatively charged) emulsions are used in this regard. Bitumen stabilisation is used for soils with plasticity indices (PIs) up to 25. Bitumen stabilisation can be used on sandy soils, but before you apply any bituminous product as a stabilising agent, you need to check with the manufacturers regarding specification and suitability.

1.9 Slopes

A slope in engineering terms is any inclined surface to the horizontal. When you drive on a road that has been cut through a hill or mountain or even where a road has been constructed on fill material, you will encounter slopes. Although they may appear insignificant, it is very important that just as much attention be paid to these areas as to the rest of the project.

A fairly common engineering failure is **slipping** of an embankment or cutting, and considerable research has been carried out into the causes of such failures. Water is frequently the cause of earth slips, either by eroding a sand stratum, lubricating a shale or increasing the moisture content of clay and hence decreasing the shear strength. When a slip in a clay soil occurs, it is frequently found to be along a circular arc and therefore this shape is assumed when studying the stability of a slope. This circular arc may cut the face of the slope, pass through the toe or be deep-seated and cause heave at the base.

Protection of the slope against environmental conditions such as wind, rain and even animals is important. Exposure to climatic conditions can increase the rate of soil erosion and hence instability of the slope. If the wind can move sand dunes in the desert then imagine what it does to an unprotected slope. Have you ever stood on the beach when the wind is blowing very strongly and felt the force of the sand grains when they strike your body? Wind like this could loosen other grains in an unprotected slope and then slowly result in disintegration. Rain, similarly, can cause just as much damage.

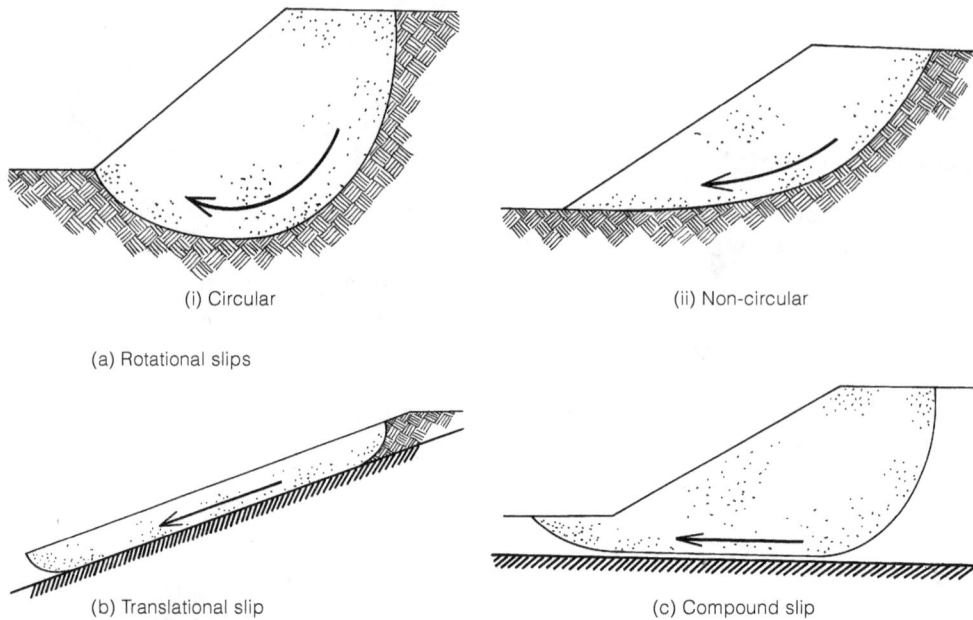

(i) Circular   (ii) Non-circular

(a) Rotational slips

(b) Translational slip   (c) Compound slip

Figure 1.40 Slope failures

Predicting the most likely failure plane relies heavily on experience, based on the study of past cases.

The cause of slope failure in a cutting will be quite different to that in an embankment. A cutting is an example of the relief of stress in the soil and is called an unloading relief of stress in the soil. However, pore pressure build-up can also occur in cuttings. The soil resistance dissipates each time and a part of the engineer's problem will be to predict the soil properties during the design life of a cutting.

Embankments and soil heaps, on the other hand, are loading cases and the construction period is the most critical time, owing to the build-up of pore pressures during construction, with the subsequent reduction in effective stress. In time, these excess pore pressures dissipate and the shear resistance of the embankment increases, although consolidation may now become the major problem.

1.9.1 Slope protection

There are various methods to protect slopes from collapse. The most important aspect of all of these is to control the moisture in the soil. Suitably designed drainage should minimise any seepage pressure and result in a more stable slope.

What would you do if told to protect a slope?

Some of the methods used in protecting slopes are:

- **Gabions:** These are large aggregates (100–300 mm) placed in a wire cage, packed on the slope, with soil material used to fill in behind these cages. Because of their weight, gabions are fairly stable and also allow for any water caught up in the soil behind the gabions to seep through
- **Hydroseeding:** Plant seeds are mixed with nutrients and fertiliser and, using a specially adapted machine, sprayed at a fairly high pressure into the soil. The pressure that the mixture is sprayed at drives the seeds into the soil
- **Rip-rap:** Mostly used on dams, rip-rap is so versatile that it can be used almost anywhere. Like gabions, it also uses very large aggregates, but the difference lies in the way that these aggregates are placed by hand, which is very time consuming and labour intensive
- **Planting:** Like rip-rap, planting is labour intensive, as it requires people to plant shrubs and plants along the slope in order to stabilise and protect it
- **Straw covering:** This is a temporary means of protecting the slope until a more permanent application is used or until the soil stabilises itself through the growing of grass or other shrubs
- **Concrete retaining block walls (CRBs):** These are another form of erosion protection of slopes, using an engineered material (like a concrete block). The use of concrete retaining blocks is relatively new in erosion protection and popular as they are versatile and can be designed to blend into the environment by allowing growth in the wall after they are placed. CRBs are often referred to by their product name, such as terraforce, but this is a mistake. A concrete retaining block must be designed by an engineer or other suitably qualified professional.

Figure 1.41 Concrete retaining block system

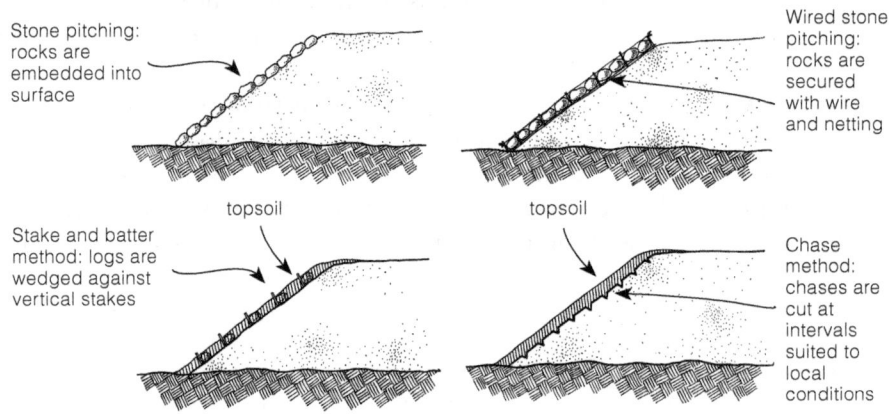

Figure 1.42 Slope protection

1.10 Reinforced earth

As the name suggests, **reinforcement** of earth relates to **strengthening** the soil to make it withstand higher demands. Reinforcement can be thought of as the same principle that is applied to concrete structures, except that in this case the reinforcement is by another material.

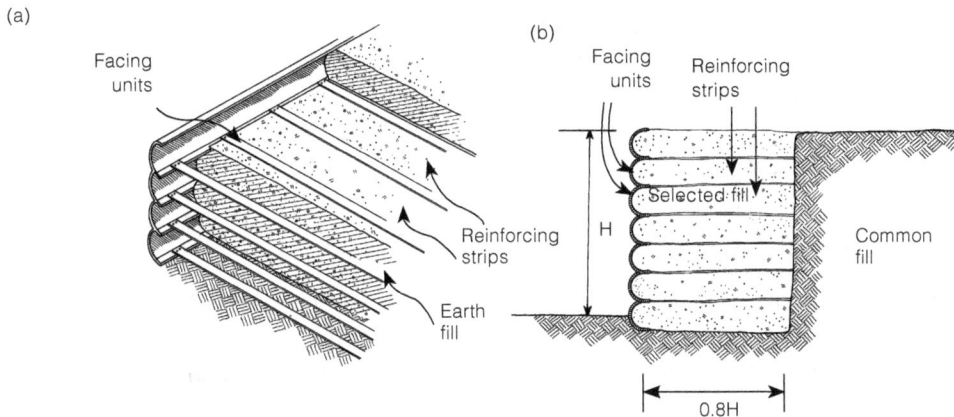

Figure 1.43 Reinforced earth

Reinforcement is any kind of linear element able to resist high traction. Practically, reinforcement can assume any shape whether strips, wires, meshes, metals, plastics or even cloth (geosynthetic material). The kind or shape of material can be selected in accordance with the type of structure to be built. It is important, however, that the type of material used is not of a kind that will disintegrate within short time frames, for example unprotected steel that will rust.

From ancient times, earth has been used as a construction material. Think back to the Egyptians using earth embankments to allow the slaves to haul those huge blocks in order to build the pyramids. Compared to other material, earth has the advantage of being cheap, but because of its poor mechanical properties, it needs to be supported in some way.

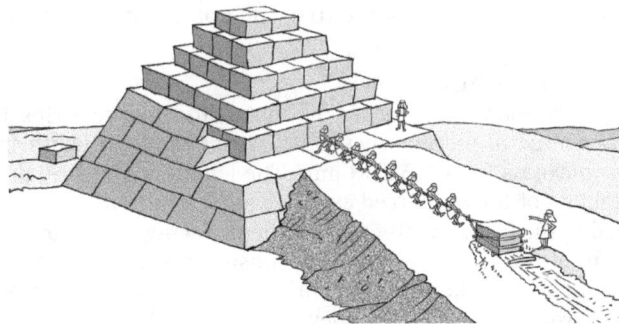

Figure 1.44 Pulling a heavy load

Reinforced earth uses the friction between the earth and reinforcement to keep the soil in place. The cost is fairly low and the mechanical properties of the soil can be improved tremendously. The soil transfers the forces built up in the earth mass to the reinforcement by means of friction. Thus, the earth will develop tension in those directions in which the reinforcement has been placed and the earth will behave as though it has cohesion. It is necessary for the fill material to have good internal friction (non-plastic material), which precludes the use of such soils as clays. It is also important to prevent the soil from running out between the reinforcing strips and for this purpose a covering or skin is necessary.

1.10.1 Application

The use of reinforced earth is generally determined by technical or economic considerations. It is one of the more versatile materials around. The most common applications of reinforced earth are:

- Walls
- Ground slabs
- Dams
- Foundations
- Embankments.

1.10.2 Methods of construction

Reinforced earth is built in stages, each stage consisting of the assembly of a new layer of facing elements and the placing of the corresponding earth fill, followed by a new layer of reinforcing strips.

Certain minimum criteria with respect to grading and water content should be met in order to ensure development of sufficient friction between the earth and reinforcement (see sections 1.3.3 and 1.3.7 on moisture content and grading and their effects). The material must not be too large, resulting in cumbersome placing or damage, or too fine causing material to constantly run out from between the layers. Water on the other hand will cause the corrosion of metal strips.

Reinforcement

Reinforcement can consist of metal and plastic strips but these need to be galvanised to prevent rust. These strips range in width from 40–120 mm and in thickness from 1.5–3.0 mm. The length of the strip depends on the internal stability of the reinforced earth mass. The one end of these strips will usually be tied to the facing of the embankment. **Facing** is the covering used to protect the front of the soil mass. This can consist of any material, but concrete panels are mostly used. When using metal strips, it is usually estimated that, if galvanised and properly constructed, they should last as long as 30 years inside the soil mass.

Facing

The purpose of the facing is to retain the soil between the layers of reinforcement in the immediate vicinity of the wall facing. The facing must be able to adapt to deformations without distortion. In many cases, the facing also doubles up as an aesthetic feature within the earth mass. Metal facings are approximately 120 kg and can be moved manually, whereas concrete ones are a lot heavier and require a small crane to move and put in place. More manageable concrete facings are 'Löffelstein' consisting of small concrete blocks that are packed on top of each other. You will normally see this type of facing in a residential area where there are steep slopes, for example at seaside developments.

Compaction

Compaction is necessary whenever there is a need to minimise settlement within the structure, for example to support a highway or carry concentrated loads. Structures such as walls for gardens and terraces, and protective walls are usually built without compaction.

There is no scaffolding required as the work is carried out on the embankment side of the structure. In other words, the side requiring reinforcing is the side you are working on. When using compaction equipment, you must ensure that you remain approximately 2 m away from the edge of the embankment, which must be compacted using lighter compaction plant, for example a BOMAG vibrating plate.

The main advantages of reinforced earth are that it is:

- A simple material, quick and easy to make
- A flexible material, able to withstand important deformations without damage
- An economical material. For big structures, the difference in cost compared to a structurally reinforced concrete wall is significant
- A heavy material when compared to concrete structures. In all material used nowadays, there exists practically only one heavy material, earth itself. It is used more and more (dams, road embankments, sea dikes, etc.), but we are compelled to let it spread over large spaces, with gentle slopes
- Technically sound and can be made to look architecturally pleasing.

Self-evaluation 1.3

1. Complete the sentences:
   a. During stabilisation, the strength of the material is _____.

   b. Stabilisation by consolidation means the _____ of a soil.

   c. _____ can cause severe burns when it comes into contact with moist skin.

d.   _____ is effective in stabilising medium to low plasticity materials.

e.   A fairly common engineering failure is _____ of a soil embankment.

f.   The essential feature of reinforced earth is the _____ between the soil and the reinforcement.

g.   Reinforced earth is a _____ material, able to withstand important deformations without damage.

2. State whether the following statements are **true** or **false**:

a.   Soil stabilisation improves the strength of the soil.

b.   Soil compaction reduces the air in the soil thereby increasing the voids.

c.   Chlorides used to stabilise soils are calcium chloride and sodium choride.

d.   One will always find cracks in a cement-treated soil layer.

e.   Rip-rap is large aggregate that is packed in a wire basket.

f.   Bitumen stabilisation is used for soils with a PI up to 25.

g.   Facing is the covering used to protect the front of the soil mass.

1.11 Summary

The purpose of this unit was to:

- Introduce you to the various applications of soil – a material that is usually taken for granted.
- Explain soil structure and its principles
- Discuss the importance of moisture in the soil
- Identify and perform some of the important tests relating to laboratory and field identification
- Identify soil characteristics when exposed to field identification
- Discuss compaction of soil and its importance
- Review the procedures followed when crushing rock
- Explain the purpose of stabilising soils using different methods
- Explain the importance of providing good protection to sloped surfaces
- Define the concept of reinforced earth.

Answers

Self-evaluation 1.1

1. a.   Topsoil

b.   Moisture and air voids

c.   Cohesive

d.   Resistance to flow

e. 25
f. Plastic limit
g. Bank, loose and compacted
2. a. True
b. True
c. False, dilatancy refers to moisture content or wetness
d. False, usually 30–40 kg are used
e. True
f. False, approximately 3 g are used
g. True
3. a. Convert Bm³ to Lm³

$\therefore 76.5 \times 1.12 = 85.7 \text{ m}^3$

$$\text{Base } \varnothing = \left[\frac{7.64V}{\tan R}\right]^{\frac{1}{3}}$$

$$= \frac{7.64 \times 85.7}{\tan 32°}$$

$$= 10.16 \text{ m}$$

$$\text{Height} = \frac{D}{2} \times \tan R$$

$$= \frac{10.16}{2} \times \tan R$$

$$= 3.1 \text{ m}$$

b.

Sieve aperture (mm)	Mass of soil retained	% soil retained	% soil passing
63.0		$(\frac{35.3}{1\,350.6} \times 100) = 2.6$	$(100 - 2.6) = 97.4$
53.0		13.7	83.7
37.5		21.3	62.4
26.5		12.8	49.6
19.0		7.7	41.9
13.2		7.0	34.9
4.75		7.3	27.6
2.0		5.4	22.2
0.425		8.6	13.6
< 0.425		13.6	0
Total	1 350.6	100	

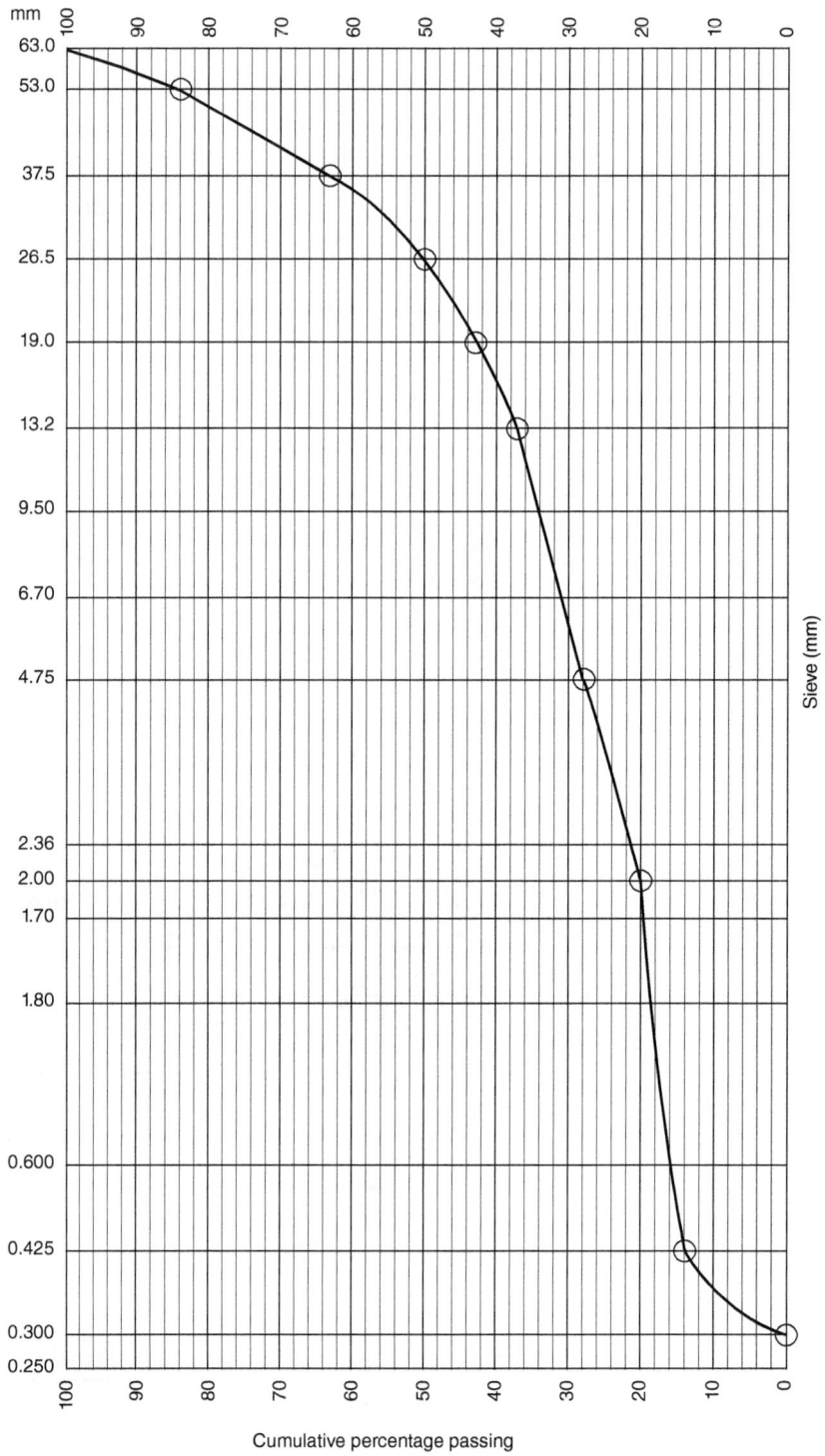

Sieve (mm)

Cumulative percentage passing

Comments on soil grading

Compare this grading curve to the shapes of those explained in example 1.2 (fig 1.18). You can see that this curve more closely resembles the shapes of a well-graded material. It would therefore be safe to say that this material is of a well-graded type.

Self-evaluation 1.2

1. a. Identify
    b. Class
    c. Compaction
    d. Quarries
    e. 1.9
    f. Stemming
    g. Misfire
2. a. False, tests are done on fine-grained soils
    b. False, specialised personnel and equipment are required for some of the tests
    c. True
    d. True
    e. True
    f. False, influenced by blasting
    g. False, the opposite is true
3. a. See section 1.6.2
    b. Refer to section 1.7.1. Any five of the following will do:
       * Design and maintenance
       * Slope or gradient
       * Drainage
       * Smoothness of road surface
       * No sharp curves
       * Level loading area
       * Gradient not more than 1.9
    c. It is a requirement of law, but also for safety and repairing damage to the environment
    d. Shock or heave energy. See section 1.7.2
    e. Usually two sensations are encountered, namely vibration and noise. See section 1.7.2
    f. It may still be unsafe due to unexploded charges or misfire
    g. See table 1.8 and section 1.7.5

Self-evaluation 1.3

1. a. Increased
    b. Pre-loading
    c. Unslaked lime
    d. Cement

  e. Slipping
  f. Friction
  g. Flexible
2. a. True
  b. False, both reduction in air as well as voids
  c. True
  d. True
  e. False, gabions are large aggregates in a wire basket
  f. True
  g. True

Advanced exercises

1. Why is it so important that we understand the properties of soil?
2. Why can we not use topsoil as an 'engineering material'?
3. What does the plasticity index (PI) of a soil mean?
4. If one tested soil sample has a PI of 6 and another has a PI of 30, what is the significance of these results and what do they indicate?
5. If a soil does not have a plasticity index, what does it mean?
6. Why is the amount of water (moisture content) in a soil of significance?
7. How important is the amount of voids in a soil?
8. If a soil derives strength from two sources, will it still function in the absence of either and why? If your answer is yes, give examples of where this can apply.
9. Why is the strength of a soil such an important factor?
10. In what situation can you use a soil that has a grading analysis of:
 a. Continuously graded?
 b. Gap graded?
 c. Open graded?
11. State whether the following statements are **true** or **false**:
 a. Plastic characteristics demonstrate the ability to display resistance to flow.
 b. LL (longitudinal limit) is the moisture content at the boundary between liquid limit and semi-solid limit.
 b. Consistency of the soil can be expressed using three defined limits.
 d. Soil classification systems allow the engineer to classify soils only if they are fine materials.
 e. Loose condition (Lm^3) is the reduced volume after compaction of soil.
 f. Stabilisation is a process to improve the condition of the soil.
12. How would you describe a soil structure?

13. Why, from an engineering point of view, is it essential to know the process of compaction in soil?
14. Soil can be described using six field observations. Name and describe these characteristics.
15. There are three different stages describing the volume of soil. What are they?
16. Mention three advantages of soil stabilisation
17. Name and explain three types of soil stabilisation.
18. Below is information extracted from a representative soil sample. You are to determine the moisture content as well as describe the importance of moisture content for a construction material.
    a.   Mass of container = 180 g
    b.   Mass of wet soil + container = 3 250 g
    c.   Mass of dry soil + container = 2 100 g

Chapter 2

Concrete

Learning outcomes

After studying this unit, you should be able to:

- Explain the basic ingredients of concrete and its characteristics
- Assess the important properties of concrete both in the fresh and hardened stage
- Identify the important factors which influence the properties of concrete
- Explain and perform some of the tests carried out on concrete
- Discuss the water:cement ratio and its influence on concrete properties
- Review the basic principles underlying concrete mix design
- Describe the making, types and uses of cement and their properties
- Explain proper storage and handling procedures for cement
- Identify the various applications of concrete.

2.1 Introduction

What is concrete?

Concrete is a material made up of cement, water, sand (fine aggregate) and stone (coarse aggregate). A mixture of cement, water and sand is called **mortar** and is used for masonry joints and plastering. A mixture of cement and water is called paste and it is this **paste** that gives concrete its strength.

Concrete has many **applications** as a construction material within the civil engineering industry, for example foundations, columns, beams, slabs, bridges, etc. Some applications are tailored to a specific purpose, as shown in table 2.1.

Table 2.1 Applications of concrete in civil engineering projects

Use	Requirement
Dams	Watertightness
Reservoirs	Watertightness
Roads	Wear resistance
Buildings	Aesthetics (good appearance)
Harbours	Abrasion resistance, durability
Nuclear shields	High density
Road barrier rails	Impact resistance
Railway sleepers	High strength
Bridges	High strength
Airport runways	Impact and wear resistance

Many engineers believe that concrete is the ultimate, friendly building material for these reasons:

- As a finished product, concrete usually **saves energy** through its good insulation properties and its ability to act as a heat store. Concrete, when properly prepared, can protect you from natural elements such as wind, rain and temperature, and because of its dense nature it can retain heat for a long period. As an example, a concrete wall or road that has been exposed to the sun all day will remain warm long after the sun has set

(a) Concrete

(b) Mortar

(c) Paste

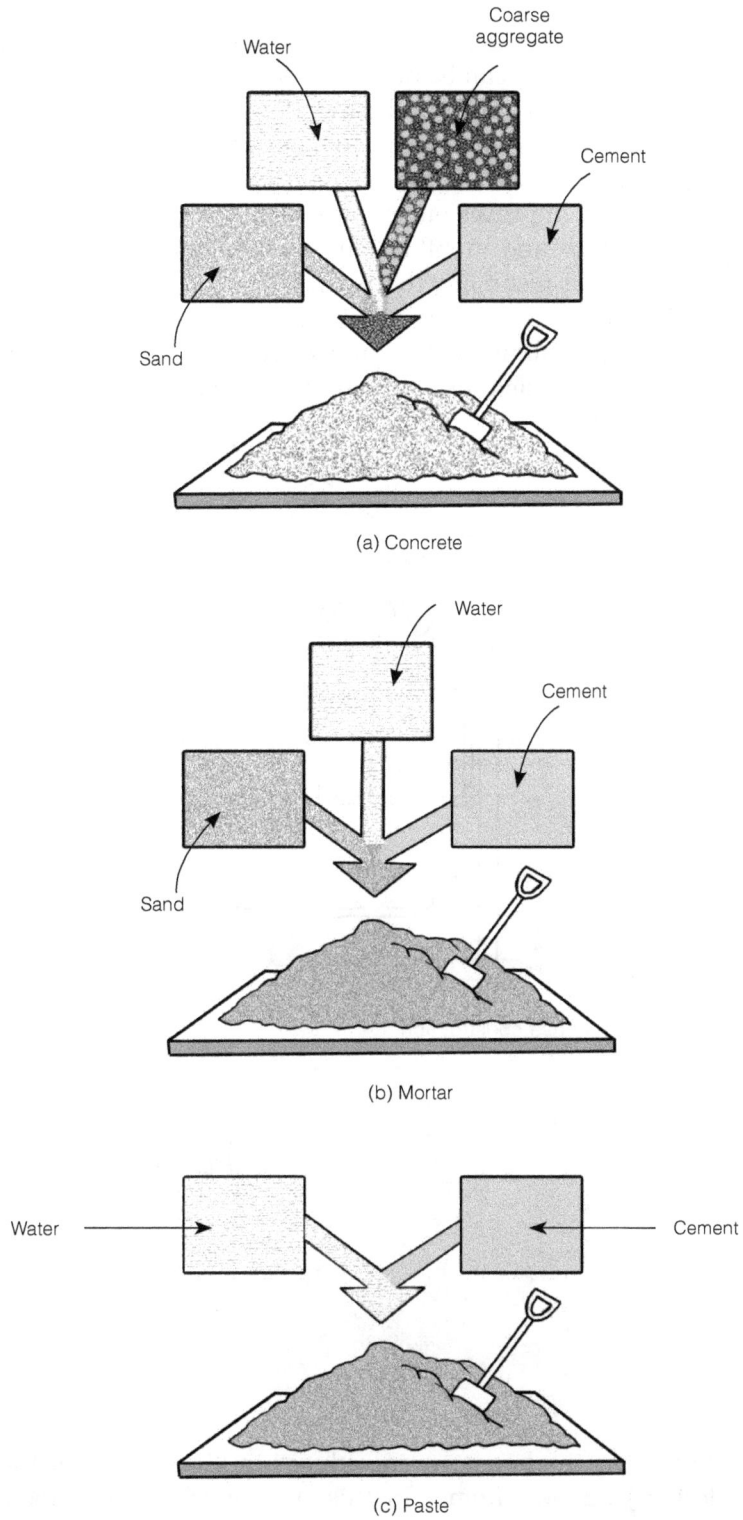

Figure 2.1 Concrete, mortar and paste

■ Concrete production does have an **effect on the environment.** However, care can be taken to make the quarrying of the limestone, clay, natural rocks and sand (for aggregate) as environmentally friendly as possible. Once a quarry is exhausted, it can be rehabilitated and given a new life, for example as a sports facility

■ Concrete improves our **quality of life** in the areas of health, safety, recreation and mobility. The world's infrastructure, building and roads bear testimony to this fact

■ Concrete's properties, such as **high compressive strength, fire resistance, mouldability, impermeability, long life** and **resistance to chemical attack,** are not easily matched by any other construction materials.

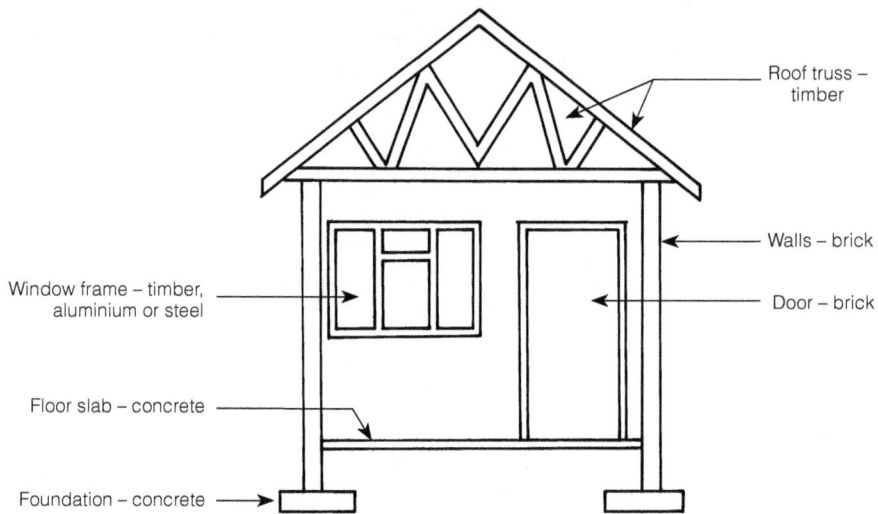

Figure 2.2 A house construction

Where in your house do you think you need cement or concrete?

Remember the house we introduced in chapter 1 and the request to design your own home? In this chapter we want to focus on where cement and concrete will be applied in the construction of the house.

As identified in the figure above, concrete is used in the foundations and the floor slab of the house, but it could also be used in other elements such as beams, columns and slabs. Mortar is used when laying the bricks and it can also be used to 'plaster' the inside and outside walls of the building to improve its appearance. If you have decided to build your house from face brick, you will not need to plaster the outside walls but certainly the inside.

2.2 General concrete specifications

Because of varying strength requirements and the way in which concrete is transported, placed and compacted, not all concrete mixtures are the same. Usually, the engineer specifies the strength while the contractor decides on the consistency or workability of the mix.

2.2.1 Mix requirements

A mix for a given strength can be specified in one of two ways. Proportions or quantities of each material to be used may be stated in terms of either **volume** or **mass**. Alternatively, a **strength requirement** may be given.

In addition, the consistency of the concrete must be specified.

Specifying proportions by volume is done by describing a mix as 1:3:3, 1:2:2, etc., meaning the ratio of cement to sand to stone. For example, 1:3:3 means one part cement to three parts sand to three parts stone (always in this order).

$$1 \quad : \quad 3 \quad : \quad 3$$
$$\downarrow \qquad \downarrow \qquad \downarrow$$

1 volume cement:3 volumes sand:3 volumes stone

On site, cement may be measured in bags and aggregates in builder's wheelbarrows. The volume of a builder's wheelbarrow is approximately equal to two bags of cement.

When specifying proportions by **mass** or **volume**, an engineer must be certain that, if the mix requirements are met, the concrete will be **strong enough**.

An example of a similar mix proportion by mass would be:

Cement	50 kg
Sand	150 kg
Stone	150 kg
Water	35 litres

If concrete is specified by strength, the strength of concrete is measured in **megapascals (MPa)**, which describes the pressure the concrete can withstand in a standard test. Different strengths are required for different uses, as shown in table 2.2.

Table 2.2 Different concrete strengths

Use	Strength
Low-strength concrete	5 MPa to 15 MPa, e.g. footings and mass concrete
Medium-strength concrete	20 MPa to 40 MPa, e.g. beams, foundations
High-strength concrete	> 50 MPa, e.g. railway sleepers, high stress members

2.2.2 Consistency

The consistency of the mix affects how easily the concrete can be placed and compacted. Depending on where the concrete is to be placed and how it is to be compacted, concrete may range from stiff to sloppy or liquid, but still have the same strength if the mix is correctly proportioned. Examples of the uses when differing consistency characteristics are required are shown in table 2.3.

Table 2.3 Concrete consistency characteristics

Consistency	Use
Stiff concrete	Dam construction
Normal concrete	General construction work
Sloppy concrete	Self-compacting/self-levelling concrete, e.g. some floors and shaft linings

2.2.3 Properties of fresh concrete

When concrete is newly made, it is a wet, plastic material and in this condition it is known as **fresh** or **plastic** concrete. Concrete is in the fresh state while its shape can be changed and it can be compacted. Concrete remains in this state for **two to three hours**. During this time, a chemical reaction between the cement and water takes place, called hydration (see section 2.3.1). Gradually the concrete stiffens, sets and gains strength. The length of this period depends on various factors including **temperature, cement type and content,** and **admixture type and content.**

Temperature

Concrete strength → Cement type and content

Admixture type and content

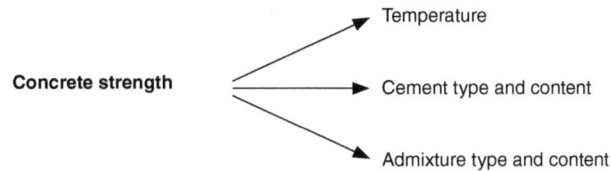

Figure 2.3 Factors influencing the hydration of concrete

While the concrete is in a fresh or plastic state, it has to be moved or transported from the mixer, placed in the formwork, compacted to remove all air and given a smooth surface and finish. There are several ways of moving the mixed concrete from the mixer to the place where it is needed, for example dumpers, chutes and conveyor belts.

Definitions **Formwork** is panels constructed – usually from wood and/or steel – to shape the concrete and hold it in place while the concrete sets and gains strength. Formwork is discussed in more detail in *Construction Methods for Civil Engineering.*

Figure 2.4 Example of column formwork

Workability

The term **workability** is used to describe the ease with which operations such as placing concrete can be carried out, using the available equipment. Even a stiff concrete can be workable when sufficient compaction energy is applied to it. Unfortunately, workability cannot be measured directly.

Workability comprises several properties, the more important ones being:
- Consistency
- Cohesiveness
- Mobility
- Compactability.

The workability of a concrete mix can be affected by **balancing** the following factors:
- **Stone size:** Smaller stone improves the **workability** but increases the cost for a given strength
- **Fines content of sand:** A lower fines content increases the cohesiveness of the concrete, while too much fines tends to make the concrete 'sticky'. 'Fines' are very small soil particles (passing a 0.075 mm or 75 μm sieve)
- **Cement content:** High cement content makes for sticky mixes
- **Stone content:** The higher the amount of stone in the mix, the harsher and more difficult it is to mix and compact. When using too little stone, the mix becomes **uneconomical** because more cement is used.

Consistency

Consistency refers to the **stiffness** or **sloppiness** of fresh concrete and is most commonly measured by a slump test. Concretes of the same consistency may differ in workability.

A slump test measures the consistency of the mix and is used to control the amount of water in a mix. The sloppier the concrete the greater (higher) the slump. The test must be carried out in accordance with SANS method 58621:2006.

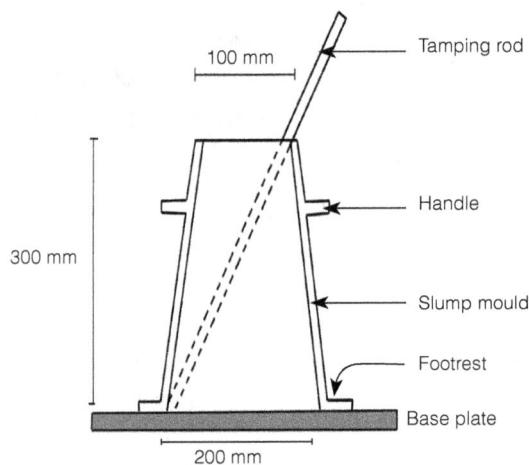

Figure 2.5 Slump test apparatus

The materials and tools for a slump test are:

- A standard mould – make sure it is clean and damp
- A tamping rod – one or both ends must be rounded
- A non-absorbent, rigid surface, e.g. a steel base plate, which must rest flat and firmly on the ground.

How is slump tested?

Start by placing the mould in the centre of the base plate. Place your feet on the foot pieces to hold the mould firmly in position. Fill the mould in **three** layers of approximately equal height. Each layer is given **25 blows** of the rounded end of the tamping rod. The blows should be made uniformly over the area of each layer. The bottom layer is tamped throughout its depth. When tamping the other layers, the rod must just penetrate the layer below. The last layer should slightly overfill the mould.

After tamping the top layer, level the surface by removing all excess concrete using a sawing and rolling action of the tamping rod. Also clean off any concrete on the mould, especially round the base.

Transfer pressure from the foot pieces to the handles. Remove the mould slowly and steadily from the concrete by raising it vertically; take a few seconds (5–10) for this. Turn the cone upside down and place it next to the slumped concrete. Place the tamping rod across the top of the mould and measure the slump to the nearest 5 mm.

The slump is the height from the bottom of the rod to the highest point of the concrete specimen.

Different types of slump are shown in fig 2.6. The test must be repeated if a shear or collapsed slump occurs. When you measure the slump, record the following information:

- Date and time
- Sample number
- Place of test
- Type of slump
- Name of tester.

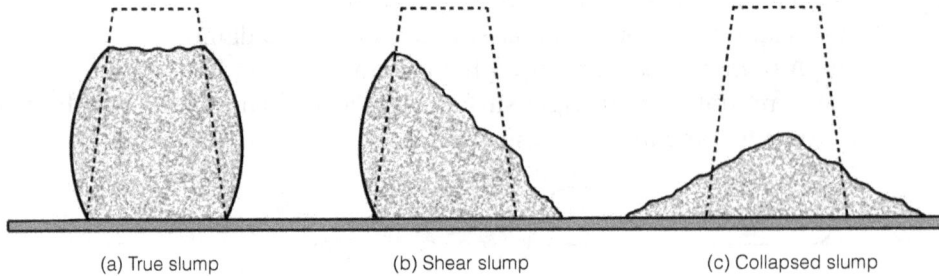

(a) True slump   (b) Shear slump   (c) Collapsed slump

Figure 2.6 Different types of slump

Concrete normally has a slump value specified. The concrete is acceptable provided the measured slump is within 25 mm or one-third of the specified value, whichever is the greatest. As an example, if the engineer specifies a slump of 75 mm, the actual test results should be between 50 and 100 mm. This figure is obtained by subtracting 25 mm ($75 \times 1/3$) from 75 mm and adding 25 mm to 75 mm.

After the test has been done, all equipment must be thoroughly cleaned with water.

EXAMPLE 1

The engineer has specified the slump of a concrete pour to a reinforced slab to be 100 mm. Calculate the upper and lower limits that the slump is allowed.

Solution

Look back to the paragraph where the greater value of 25 mm or 1/3 of the specified slump is to be used. Therefore 100 mm × 0.333 = 33 mm, which is greater than 25 mm. In other words, the limits must be:
Upper: 100 mm +33 = 133 mm
Lower: 100 mm – 33 = 67 mm

What do you do after retesting twice if the results are not within the accepted standard?

If the concrete is too stiff, you can ask the supplier to add more water, then mix and retest. However, if this concrete is still too wet it may have to be returned and a fresh mix used

2.2.4 The slump test

Equipment

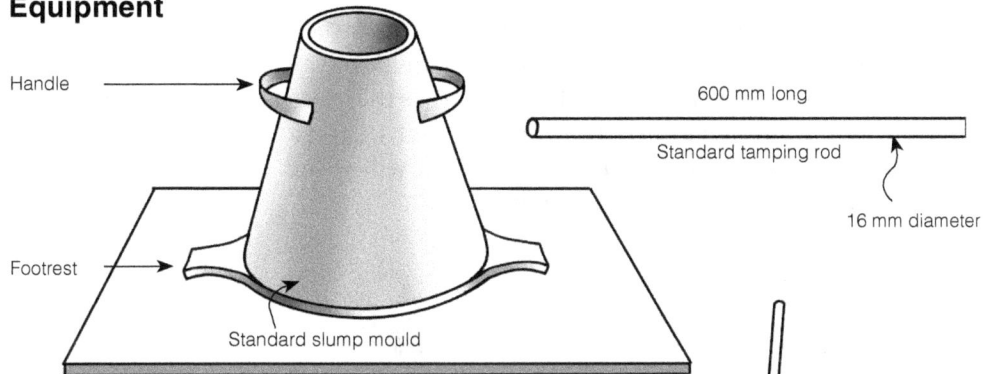

Handle

600 mm long

Standard tamping rod

16 mm diameter

Footrest

Standard slump mould

Method

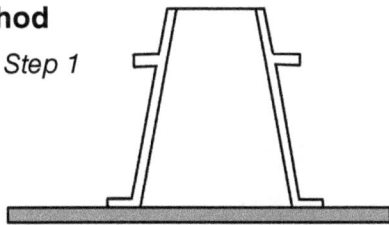

Step 1

Smooth, hard surface

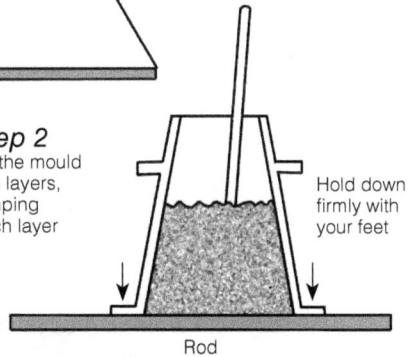

Step 2
Fill the mould in 3 layers, tamping each layer

Hold down firmly with your feet

Rod

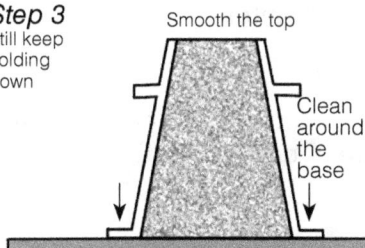

Step 3
Still keep holding down

Smooth the top

Clean around the base

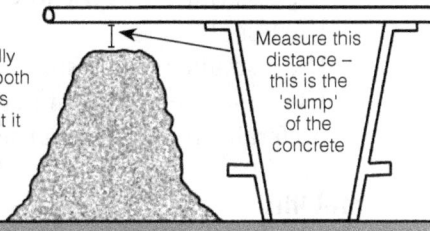

Step 4
Lift the mould vertically using both handles and put it down beside the mould

Measure this distance – this is the 'slump' of the concrete

Three kinds of slump may appear

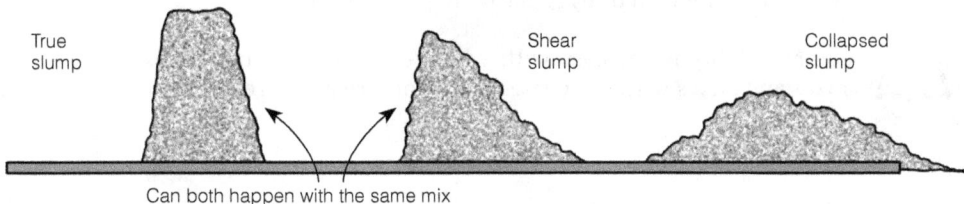

True slump

Shear slump

Collapsed slump

Can both happen with the same mix

Figure 2.7 Concept map of the slump test

Cohesiveness

Cohesiveness describes the ability of concrete to remain well mixed. When all the ingredients in concrete stick together, the concrete is cohesive. A sticky mix is highly cohesive. If the mix is either too wet or too stony, the stone will easily separate from the mortar, resulting in segregation which could mean incomplete compaction or honeycombed concrete. **Segregation** may occur when transporting wet concrete in dumpers. The stone settles to the bottom and water rises to the surface or it may occur when concrete is discharged from a chute.

If concrete is allowed to segregate, the mixture will not be uniform when placed in the structure. Some areas will have a high mortar content, others will be very stony. In stony areas there may not be enough mortar to fill the spaces between the stones. If there is insufficient mortar, the concrete will be **honeycombed**. **This will result in the concrete being weak and could cause it to fail.**

How can cohesiveness be improved or segregation reduced?

To improve cohesiveness or reduce segregation, one or more of the following steps can be taken:
- Use less water (i.e. use lower slump concrete)
- Use less stone
- Use a smaller size of stone
- Use more very fine material (the use of a fine blending sand is preferable to increasing the amount of original sand).

Mobility

Mobility is the ease with which concrete will flow around obstructions, such as reinforcement, and into corners of formwork. Highly mobile mixes are required in heavily reinforced or irregularly shaped sections.

Compactability is the ease with which entrapped air can be expelled. Stiffer mixes and mixes with large-sized aggregates require greater effort to compact.

When concrete has been placed and compacted, some matter will migrate to the surface to form a layer of water. This is known as **bleeding** *or* **water gain**.

Bleeding occurs as the relatively heavy solids (cement and aggregate particles) settle downwards. When the cement paste stiffens, bleeding stops.

Some bleed water may get trapped under aggregate particles and/or reinforcement and reduce the bond or grip between the mortar and stone, which reduces the strength of the concrete.

The most common causes of bleeding are excessive water content or a lack of fine material in the sand.

How can bleeding be reduced?

Bleeding of concrete may be **reduced** by one or more of the following:

■ Use less water (lower slump)
■ Use more very fine material
■ Increasing the cementitious content (expensive).

Plastic cracking

Two types of cracking in concrete may occur while it is still plastic:

■ Plastic shrinkage
■ Plastic settlement cracks.

Plastic shrinkage cracks result from a **rapid loss of moisture** from the exposed concrete surface due to evaporation, or from the bottom surface into an absorbent material such as soil. They are most common on slabs, frequently passing right through the slab, and tend to run parallel, spaced about 200–600 mm apart.

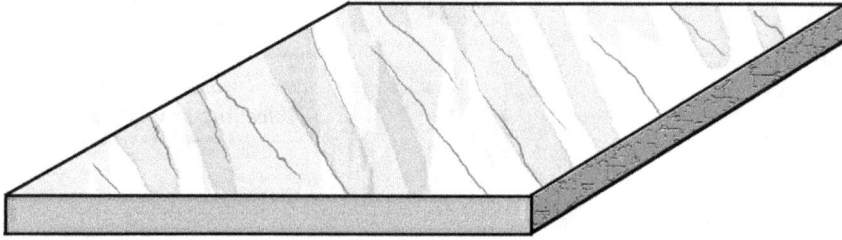

Figure 2.8 Plastic shrinkage cracks

Plastic settlement cracks occur if the settlement of solid particles, which accompanies bleeding, is restrained. Restraint may be provided by reinforcement, especially top bars in beams, slabs and stirrups in columns, and changes in direction of formwork. Settlement cracks may be minimised by reducing bleeding, or by revibration after most of the settlement has taken place, but before the concrete has set.

Hardening concrete

As the concrete loses water, it begins to stiffen. Moisture loss is due at first to loss of water into the atmosphere (evaporation) or formwork (absorption). The rate of stiffening will depend on the cement type and content, temperature, humidity, etc. Thereafter, stiffening and hardening are due to hydration (the chemical reaction between cement and water). Provided the concrete does not **freeze,** it will set, harden and gain in strength.

Medium-strength concrete can be expected to set in 34 hours in **normal weather.**

2.2.5 Properties of concrete at an early age

As concrete sets, it changes into a weak solid and then gradually gets stronger. At the same time, the concrete gets warmer and then gradually cools. A term used to describe concrete soon after it has set but is still weak is **green concrete.**

While concrete is in the green state, it can easily be damaged and it may be necessary to protect it from physical harm. The surface can, however, be worked without affecting the rest of the concrete. Trowelling or wire brushing can be done while the concrete is in this state, but timing is crucial because the strength and hardness of the surface change rapidly.

The strength and hardness of concrete are the result of chemical reactions between cement, water and cement extenders, if used. These reactions produce heat which causes increases in temperature of the concrete.

Like virtually all material, concrete **expands** as it gets warmer and contracts as it cools. Expansions and contractions are proportional to temperature increase and decrease respectively.

To give you some idea of the extent of such movements, a piece of concrete 1 m long would expand by about 0.3 mm if its temperature were increased by 30 °C.

Imagine that a 1 m long piece of concrete is clamped firmly at the ends while it is in a warm, expanded condition, and then allowed to cool through 30 °C. The contraction, restrained by the clamps, would result in stresses high enough to crack the concrete. Once the concrete has cooled completely, it would have a crack 0.3 mm wide, right through it, called a **shrinkage crack**. Shrinkage cracks are not only caused by temperature changes but by volume due to drying.

In a real structure, the movement of concrete can be restrained by adjoining concrete.

It is also possible for the heat of cementing reactions to cause temperature differences within the concrete, high enough to result in cracking. This is known as **thermal cracking**.

Cracking can be reduced by providing contraction joints.

2.2.6 Properties of hardened concrete

Compressive strength is usually considered to be the **most important property** of hardened concrete. Other requirements are that the concrete will appear **acceptable** and last for the designed life under expected conditions of use (durability).

The main factor which affects the strength of concrete is the **water:cement ratio (w:c)**. This is stated as a decimal fraction. For example, with one part of water to two parts of cement, w:c = 0.5.

Compressive strength is the resistance of standard, plane-ended specimens of concrete to crushing. Strength is expressed as a stress which is force per unit area on which the force acts, i.e. strength = F/A.

Compressive strength is measured by the cube test as described in the following standard test methods:

- Sampling of concrete: SANS method 5861 2: 2006
- Making and curing cubes: SANS method 5861 3: 2006
- Crushing cubes: SANS method 5863: 2006.

The cube test

Three cubes are needed for **each test** to obtain an average result. If we test concrete cubes at ages 7 and 28 days (as is normal), you will need six cubes – three for each test.

Table 2.4 Materials and tools

Materials	Tools
A sample of concrete (about half a wheelbarrow full)	A steel tamping rod, 600 mm long with a diameter of 16 mm that has at least one end rounded
Three standard moulds (150 mm x 150 mm x 150 mm) for each test	A wheelbarrow and a shovel
Pieces of writing paper (absorbent paper) for labels	A ballpoint pen or, preferably, a soft pencil
Mould release oil	A scoop
Grease	A steel float

When sampling from a ready-mix truck, the concrete must be divided into equal parts (usually about four) and a spade or scoopful must be taken from each part. Be sure not to take a sample from the first or last 0.5 m³ of the truck.

When taking samples from a heap of freshly made concrete, distribute the heap into four equal parts and follow the same procedure by taking enough from each part for your test.

Take sample from a moving stream

Figure 2.9 Sampling from a concrete truck

How to make the concrete cubes

- Check that:
  - Moulds are clean and do not have dust or dirt on them
  - The joint faces have been greased
  - They are assembled in the right way
  - The bolts are tight.
- Smear release oil very thinly on the inside faces of the moulds and place the moulds on a firm, level surface.
- Mix the concrete well in the wheelbarrow.
- Fill the moulds with concrete in **50 mm layers**, tamping each layer at least 45 times with the **rounded end** of the tamping rod to get the air bubbles out.
- The last layer should more than fill the mould. After tamping the last layer, use the steel float to strike off the surface of the concrete so that it is level with the top of the mould.
- Write down the following on a label for each cube:
  - The company's name
  - The contract number or the reference number
  - The date when the cube was made.
- Gently press the label onto the top of the cube.
- Cover the cubes with damp sacking followed by a sheet of plastic and store them in the shade, away from wind and where they will not be disturbed.
- **If the weather is cold** make and store the cubes indoors.
- **The next day,** loosen all the bolts and gently remove the sides of the mould.

Testing concrete cubes

Equipment
150 mm cast iron or
steel cube mould

To obtain uniform cube specimens for density and
crushing strength determinations

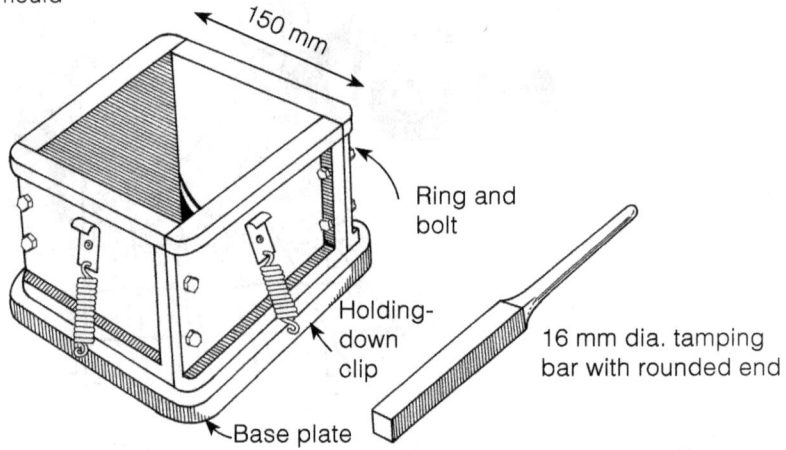

150 mm

Ring and
bolt

Holding-
down
clip

16 mm dia. tamping
bar with rounded end

Base plate

Method
Step 1
Fill the moulds
in 3 even layers

Tamp at least 45 times
per layer

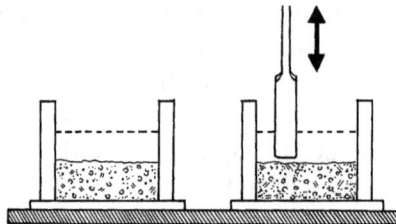

Finish off surface with a trowel
Put a paper label on each cube

Step 2

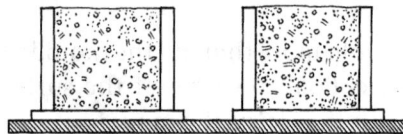

Then leave in the
mould for 24 hours
under damp sacking

Step 3

Curing
tank

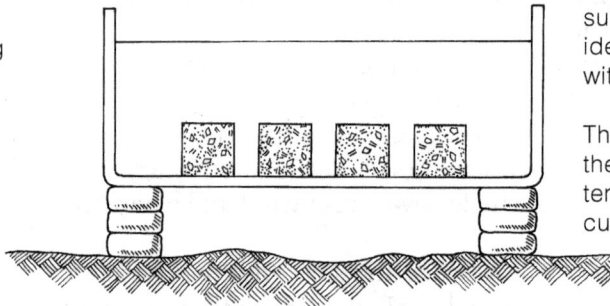

Mark the cubes
suitably for future
identification, e.g.
with a wax crayon

Then strip them and leave
them in a
temperature-controlled
curing tank

Do

Make sure that the mould faces are thinly coated with mould oil

Make sure the sections are bolted together tightly

For sampling, use a bucket or a scoop not a shovel as large pieces drop off

Take enough samples to make all the cubes required

Take the mould apart completely when removing the cube

Clean the moulds as soon as the cubes have been stripped

Make sure the curing tank water is within standard temperature limits

Company name			
Site address			
Order no.			
Quantity			
Type of material	Date of order	Date required	Signature

Keep an accurate record of all cubes taken

Don't

Don't take samples from the beginning of a mixer discharge, or the end, or from the edge of any placed concrete

YES!

NO!

fine aggregate

segregation

Don't cause segregation or accumulation of fine aggregate when compacting

Don't leave the cubes where they can be shaken or jolted before they are put into the curing tank

Don't transport the cubes until 24 hours after making

It is a good idea to write the information onto the label on the cube with a waterproof crayon.

- Put the cubes in a bath of water. The temperature of the water in the bath should be between 22 and 25 °C.
- Leave the cubes covered with water until they are taken to the testing laboratory. If the cubes have to be transported there, put them on a rubber mat, cover them with a damp sack and make sure they cannot be thrown about.
- Clean the moulds thoroughly, including the joints, lightly oil, and assemble them again.

After **7** and **28** days the cubes are tested for **compressive strength**. Any surface water, grit, etc. must be wiped off, and the mass of the cube should be determined to the nearest **10 g**. The cube is then placed centrally in the jaws (platen) of the machine with the trowelled face vertical, i.e. not in contact with either platen. A load is applied at a constant rate of 15 MPa per minute until no further load can be carried by the specimen.

The compressive strength is calculated by dividing the maximum load in **Newtons (N)** by the **cross-sectional area (A)** of the specimen. Most crushing machines give a result in kiloNewtons (kN).

The way in which a cube fails should be carefully observed. A horizontal crack (see fig 2.10) indicates that the cube was placed eccentrically in the testing machine.

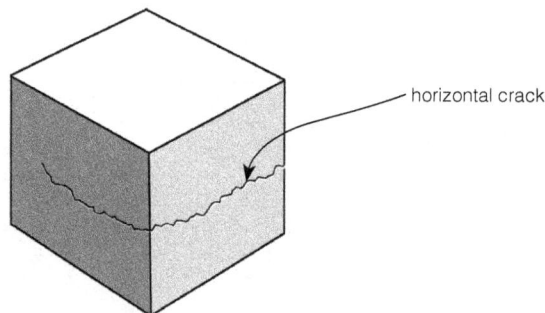

horizontal crack

Figure 2.10 Concrete cube failure

Deformation and volume change

Definitions

Material **deforms** when a force is applied to it. In some cases, the deformation is entirely **reversible**, for example when you bend an eraser, it snaps back to its original shape when you release it. This is called **elastic deformation**. In other cases, the deformation is **permanent**, for example when you bend a paper clip, it remains bent. This is called **plastic deformation**. Many materials show both behaviours – initially they deform elastically, but as the load increases, they deform plastically. To understand how concrete works, and explain its behaviour, we need to know more about changes to the shape and size of concrete. These changes are small, sometimes even too small to be seen by the naked eye, but they have important effects on the behaviour of elements and structures.

Concrete deforms as a result of the conditions in which it is placed, loads on the structure, or both of these factors.

After curing, when concrete is exposed to the elements, water will evaporate from the concrete resulting in **shrinkage** (normally between 0.3 and 0.5 mm/m). The amount of shrinkage depends on the thickness of the concrete, how dry the air is and the water content of the concrete when it was placed. If you wet dry concrete thoroughly, for example by soaking in water, it will expand, but not to its original length. Drying shrinkage manifests itself (makes itself visible) by cracking the concrete.

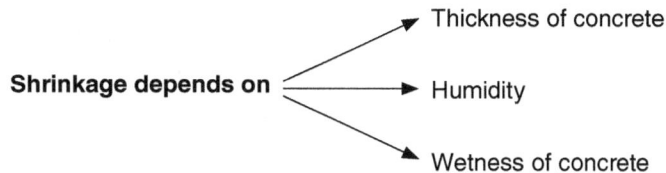

Shrinkage depends on
- Thickness of concrete
- Humidity
- Wetness of concrete

Figure 2.11 Shrinkage factors

As concrete is **loaded**, it shortens in the direction of the force. If, for example, we were to measure the height of a cube in a cube press, we would find that it gets shorter as the load increases. Just before **failure** (when the concrete cracks or, in extreme cases, shatters as it is no longer able to take a load) a typical 150 mm cube would be about 0.15 mm shorter at breaking strength.

In real structures, which may be tens of metres high or long, movements are big enough to cause problems if not designed for. Loads may be due to dead load (weight of the structure), imposed load, wind load or a combination of all three.

Dead load is the weight of the structure that is taken into account, i.e. each and every section or component of the structure, whether foundations, beams, slabs or columns, is converted to mass. **Imposed loads** are any movable items such as furniture, people, etc. that contribute to the overall mass of the structure.

Wind-loading is caused by the movement of the structure as the wind blows onto it. Imagine walking in a very strong wind and trying to resist the force of the wind. You have to compensate for this and the structural engineers need to take this into account when designing.

If concrete is loaded, it deforms immediately and then slowly continues to deform further with time. Engineers use the term 'elastic' to describe immediate deformation, and 'creep' to describe long-term deformation.

Beams and slabs will sag when loaded, whilst columns will tend to compress or shorten.

The **density** of concrete is expressed as kilograms per cubic metre (kg/m^3). The density of a specific concrete depends mainly on the density of the aggregate used, but the water and air contents of the concrete as mixed also make a difference.

Normal structural concrete has a density of between 2 250 and 2 450 kg/m^3. It must be noted that reinforcement also adds to the density of concrete.

Self-evaluation 2.1

1. Complete the sentences:
   a. A mixture of cement and water is called _____.
   b. Concrete mixes may be specified by _____, _____ or _____.
   c. The strength of concrete is measured in _____.
   d. Newly made concrete is referred to as _____ or _____ concrete.
   e. A _____ measures the consistency of fresh concrete.
   f. A condition known as _____ refers to when the stone separates from the mortar.
   g. A term used to describe concrete soon after it has set but is solid and still weak is _____.
   h. The main factor which affects the strength of concrete is the _____: _____ _____.
   i. For the cube test, moulds are to be filled in _____ mm layers and tamped _____ times with a tamping rod.

2. State whether the following statements are **true** or **false**:
   a. The strength of concrete is influenced by the type and amount of stone.
   b. All concrete mixtures are the same.
   c. Wet concrete will have a higher slump than dry concrete.
   d. When tamping the concrete during a slump test, ensure that the tamping rod penetrates to the base plate at all times.

e. Cohesiveness is the ability of the concrete to stick together.
f. The most common cause of bleeding is the existence of too much fines and too little water in the mix.
g. Environmental conditions may affect the performance of concrete.
h. The strength and hardness of concrete are a result of hydration.
i. Increasing the water content whilst the cement content remains the same, will result in a stronger concrete.

3. Answer the following:
   a. Explain how segregation can result in poor quality concrete.
   b. What do we do to concrete when we compact it?
   c. Define what concrete is.
   d. How can we improve cohesiveness of a mix without putting in extra cement?
   e. Explain the importance of the water:cement ratio in concrete.
   f. Explain the importance of workability of concrete.
   g. What is meant by consistency of a mix?
   h. What is the cause of plastic settlement cracking?
   i. Bleeding may be reduced by increasing the cement content of a mix. Why is this not recommended?
   j. When a concrete truck delivers concrete to a site, explain how a sample is obtained to do the cube test.
   k. What is the difference between a contraction joint and a construction joint?

2.3 Materials for concrete

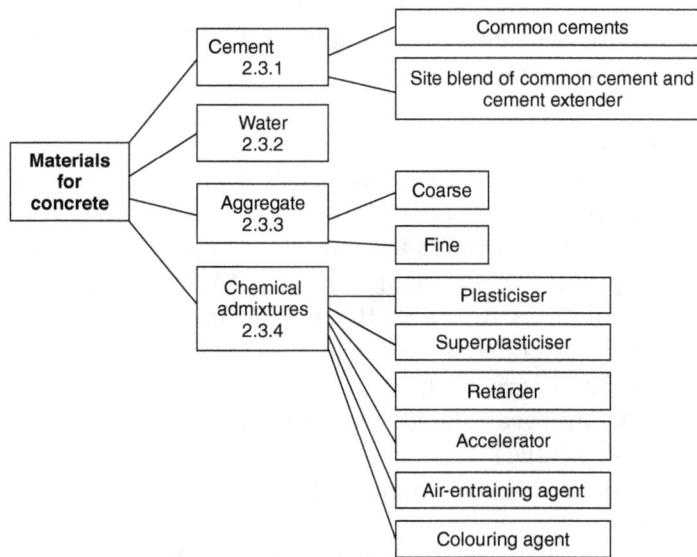

Figure 2.12 Flow diagram of the four basic materials used to make concrete

2.3.1 Cement

Evidence of the use of cement is believed to have been found in the hidden civilisation of Mohenjodaro dating back 5 000 years. The ancient Greeks used some form of mortar, but it remained for the Romans to develop and use cement.

Cement is a substance that is used in a soft or plastic state which then hardens to make things stick together. It can thus be regarded as a **binder, glue** or **adhesive** which in its hardened state binds aggregate particles to form a strong, rigid composite.

Remember the mixture of cement, sand, stone and water is referred to as concrete.

There are **various substances** which can act as cement but we will be concentrating on the specific types used for concrete within the building and construction industry.

Cements for construction are supplied in powder form (either in bags or bulk) and if mixed with water, will set and develop strength. Such cement is used extensively in most parts of the world because the raw materials are available in most regions. Also, these cements are relatively **cheap** and **versatile** as setting takes place at normal temperature and pressure, and they can be used under water, and result in a strong and durable concrete.

In South Africa, cementitious material for concrete may be either
- 'Common' cement, or
- A site blend of a common cement and a cement extender.

These types are discussed below.

Masonry cements that comply with SANS 50413-1:2007 must not be used for concrete.

Common cements
These must comply with SANS 501971:2006 *Cement composition, specifications and conformity criteria. Part 1: Common cements.*

Common cements are manufactured in five categories according to composition, designated CEM I to CEM V. CEM I consists essentially of cement on its own. CEM II to CEM V are factory blends of cement and a cement extender, or a filler such as finely ground limestone. Three types of extender are used:
- Ground granulated blast-furnace slag (GGBS)
- Fly ash (FA)
- Condensed silica fume (CSF).

Common cements are also classed according to strength, measured in a standard test at ages 2 or 7 days, and at 28 days.

Site blends of a common cement and cement extender

By site blends we mean cementitious materials mixed in the concrete mixer.

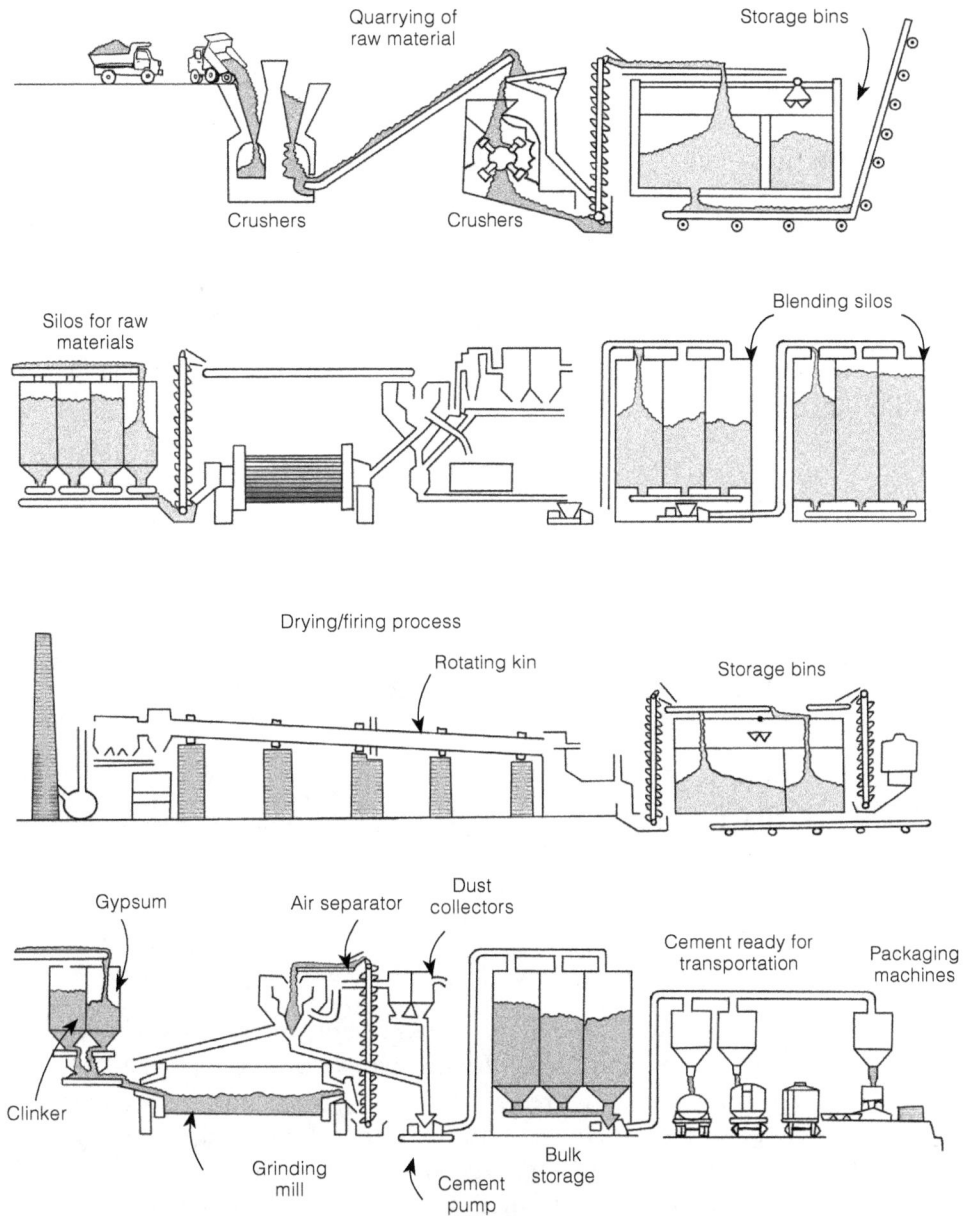

Quarrying of
raw material

Storage bins

Crushers

Crushers

Silos for raw
materials

Blending silos

Drying/firing process

Rotating kin

Storage bins

Gypsum

Air separator

Dust
collectors

Cement ready for
transportation

Packaging
machines

Clinker

Grinding
mill

Cement
pump

Bulk
storage

Figure 2.13 Manufacture of cement

Only one category of common cement is suitable for blending, namely CEM I.

Cement extenders must comply with the relevant part of SANS 1491:

- Part I: GGBS
- Part II: FA
- Part III: CSF.

The most widely used blend proportions are, by mass:

- 50% CEM I:50% GGBS
- 70% CEM I:30% FA
- 92% CEM I:8% CSF.

Raw materials used in manufacturing cement

The raw materials used in the manufacture of cement are classified into three broad categories as shown in table 2.5.

Table 2.5

Raw material category	Possible source
Calcareous: containing calcium, usually calcium carbonate	Limestone, calcrete, chalk
Argillaceous: composed mainly of clay or shale	Clay or shale
Iron oxide	Clay, shale or natural iron oxide

Activity 1

Use these materials to do the following experiments:

- cement
- water
- building sand
- concrete stone.

1. Mix a handful of cement with a cup of water in a mixing bowl. Add the water slowly, stir and observe what happens in the mixing bowl.
2. Take a handful of cement, two handfuls of sand and a cup of water and mix in a mixing bowl. Stir and note the changes.
3. Take a handful of cement, two handfuls of sand, two handfuls of stone and a cup of water and mix together in a mixing bowl. Stir and note the changes that occur.

Take note of the following:

- Is there a change in the temperature of the mix whilst mixing?

- The consistency of each mix and its workability
- The stiffness or sloppiness of the mix

Leave the mixes in a 'safe' place and check every hour to note the changes each mix undergoes, particularly in strength gain over time. Discuss your findings with your classmates.

Manufacture of cement

The raw materials are finely ground, then mixed or blended to proportions determined by the chemist. The blended material is then heated to about 1 400 °C in a kiln. A kiln is like a large oven capable of very high temperatures in which the raw materials are burnt. The material that comes out of the kiln is in the form of dark grey nodules, called clinker.

The **clinker** is finely ground with a small proportion of gypsum (2.53%) by mass of cement which stops the cement setting too quickly. **Gypsum** is a mineral composed of calcium sulphate. This blended product (normally light grey, but may also be medium grey or brownish grey) is **cement**. The colour of cement has no effect on its other properties.

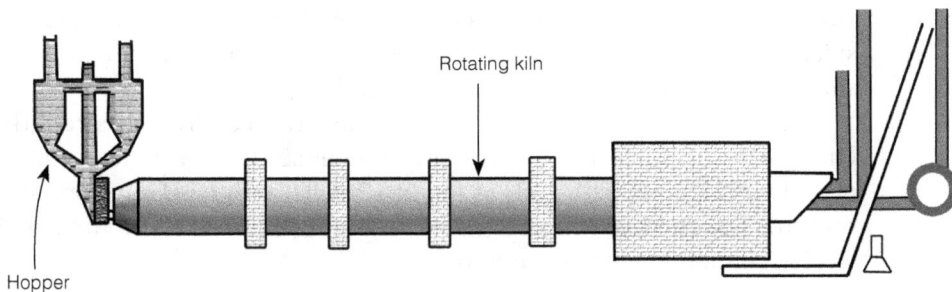

Figure 2.14 Cement kiln

Hydration of cement

When you mix cement and water, the cement particles start to dissolve slowly. A reaction, called **hydration**, takes place between cement and water and produces fibres that grow from the cement particles into the water. These fibres interlock with fibres from nearby cement particles, binding the mix together.

With time, the fibres from neighbouring particles grow together and form a strong rigid mass which holds the concrete together and gives it strength and density. During the hydration process, heat is given off. The hydration process is therefore termed **exothermic**.

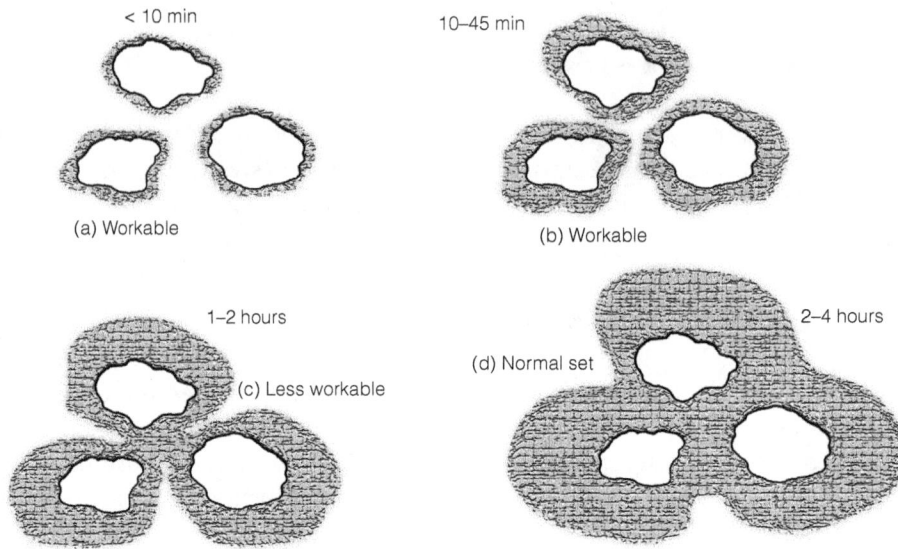

Figure 2.15 Hydration of cement

Setting and hardening

The dominant compounds in cement are **calcium silicates**. When you mix cement with water to create concrete or mortar, the calcium silicates react with the water to produce a 'gel' consisting of calcium silicate hydrate and lime (calcium hydroxide).

Gel in hardened cement paste is rigid and gives concrete its strength. Lime does not contribute to strength but causes the hardened cement paste to be alkaline with a pH of about 12.5. This high pH environment prevents the rusting of steel and so makes it possible to use uncoated steel as reinforcement in concrete.

Why should fresh concrete not come into direct contact with your skin?

Because cement paste contains lime which reacts with your skin resulting in the rapid drying and damaging of your skin.

If you mix cement and water to make a sample of cement paste, you will notice that little appears to happen for some hours. The mixture then starts to set, i.e. it changes from a plastic state to a weak solid and then hardens. Cement paste in concrete goes through the same stages.

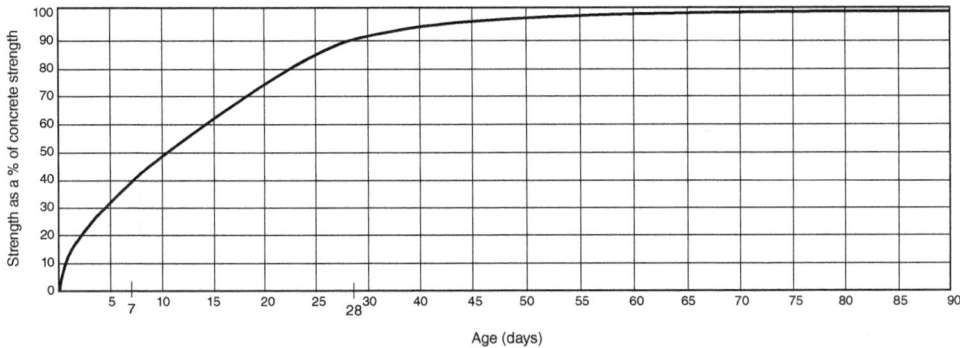

Figure 2.16 Strength gain of concrete

As you can see by looking at figure 2.16, the hardened concrete gains strength, rapidly at first but more slowly over time.

Remember that the cement will continue to hydrate, and in so doing gain strength, only while water is available. This is the reason why concrete should be cured a few days after pouring. Curing refers to a process during which the freshly poured concrete is covered or protected from drying out.

Activity 2

Curing is a very important part of the strength gain of concrete. Form into groups of about four to discuss examples of curing methods adopted on construction sites. Talk to construction personnel or have a look in a library to find out more information. Share your findings with other groups.

The long-term strength of well-cured concrete depends on the type of cement and ratio of water to cement. (Here 'cement' refers to all cementitious material.) **Typical water:cement ratios**, expressed as w:c, which are applicable and dependent on the strength requirement of the concrete are:

- High-strength concrete   0.25–0.40
- Conventional concrete   0.45–0.80
- Sand-cement floor screed 0.55–0.60

> *The setting of cement, and therefore concrete, can be influenced by certain activities. Setting time of cement can either be retarded or accelerated to suit a particular construction need.*

There are two forms of rapid setting which can occur in concrete:

- **Flash set** is caused when **too little gypsum** is used in the cement mix, resulting in the cement paste heating up excessively as it stiffens with the resultant loss in plasticity which is not restored upon remixing
- **False set** is caused by **overheating** during grinding of the clinker. Little heat is given off as the concrete stiffens. When the concrete is remixed without additional water, it regains plasticity and may be handled and placed in the normal way.

Cement extenders: sources and cementing reactions

In South Africa, cement extenders are used extensively for concrete. The main reasons for this widespread use are:

- Extenders, especially ground granulated blast-furnace slag and fly ash, reduce the material cost of concrete in most parts of the country
- Extenders improve and densify the micro structure of the hardened cement paste, resulting in greater impermeability and durability of the concrete
- Extenders can improve durability in specific cases, e.g. in marine environments and where aggregates can be attacked by alkalis.

Ground granulated blast-furnace slag (GGBS), trade name **slagment**, is a by-product of the iron-making process. The slag is rapidly chilled or quenched (causing it to become glassy) and ground to a fine powder.

When mixed with water, GGBS hydrates to form cementing compounds similar to those formed by cement (PC). The rate of hydration is, however, too slow for practical construction work unless activated by an alkaline (high pH) environment like that which PC provides. PC thus acts as an activator to speed up the hydration of GGBS. However, even activated by PC, GGBS hydrates more slowly than cement. GGBS should never be used on its own as a binder for concrete.

Fly ash (FA) is collected by electrostatic precipitators from the flues of power stations that burn pulverised (finely ground) coal. The finer fractions of the ash are used as a cement extender. In the presence of water, FA reacts with lime to form cementing compounds. This reaction is pozzolanic and FA is therefore a synthetic pozzolan.

The reaction of FA is slower than the hydration of cement. This difference affects the blends of FA and PC. The effect of FA on the properties of concrete depends on the FA content of the binder.

Condensed silica fume (CSF) is the condensed vapour by-product of the ferro-silicon smelting process and is sold in South Africa under the tradename CSF90. It is an extremely fine powder – much finer than cement. CSF is a pozzolan, i.e. it reacts with lime in the presence of water to form cementing compounds. Because the hydration of cement produces lime, pozzolans are well suited to be used with PC. CSF should never be used on its own as a binder for concrete.

Because CSF is so fine and highly reactive, these blends tend to produce higher strengths than the same amount of cement on its own. These blends are more expensive than ordinary PC and are therefore used for **specialised applications** only. Skilled personnel and the proper equipment are necessary to ensure that the proper blend ratio is achieved. Cement and CSF are mixed in ratios of 95:5 or 90:10 (PC:CSF).

Masonry cements are formulated primarily to impart good workability to mixes for plastering and joints for masonry work.

Masonry cements are a blend of cement and finely ground limestone or hydrated lime. Some masonry cements include an air-entraining agent.

The South African standard for masonry cements is SANS 504131 *Masonry cement. Part 1: Specification.* The standard specifies composition, strength performance, fineness, setting times, soundness and the properties of fresh mortar.

Masonry cements must not be used for concrete.

Storage and handling

Having decided and selected the type of cement required, it is necessary to ensure that you get what you want in terms of **quality** and **quantity**, and that quality is maintained during storage on site. Cement may be delivered to site either in **bags** or in **bulk**.

Cement must be stored in such a way that it stays dry. Because air contains moisture, contact with the air should be kept to a minimum.

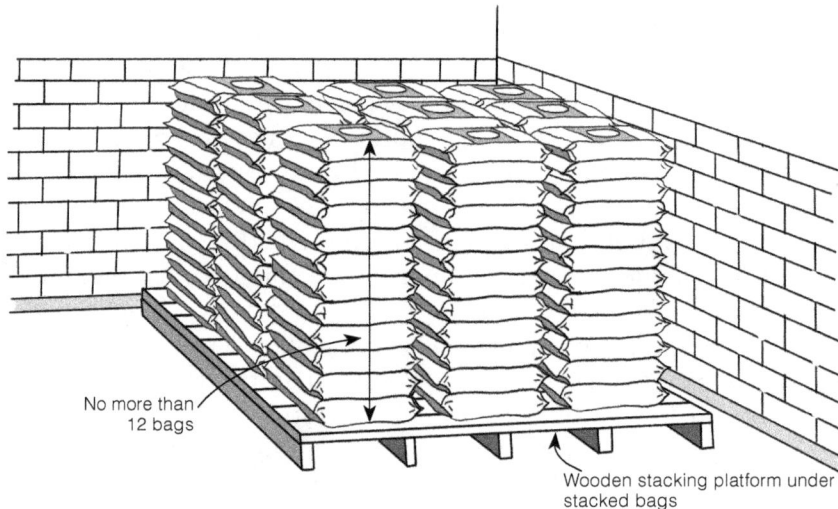

No more than 12 bags

Wooden stacking platform under stacked bags

Figure 2.17 Storage of cement in bags

It is essential that bagged cement is **stored** correctly:

- The cement store should be **weatherproof** and solidly constructed
- It should be provided with a **damp-proof floor** which should be either covered with a heavy-duty plastic sheet or raised wooden stacking platform. When stacking or transporting, ensure that there are **no sharp protrusions** such as nails, splinters, etc. which can pierce the bag
- Avoid stacking cement on corrugated iron sheets due to sharp edges and corners

- Bags of cement must be closely stacked in order to **reduce air circulation between them to a minimum**
- They should not be packed against **outside walls**
- Arrangement of the packing order should be **'first in, first out'**
- Doors should be opened as seldom as possible
- Stacks should not exceed **12 bags in height**. For estimating storage space, assume that one cubic metre of space is required to store 20 bags
- Bags of cement must **not be dropped**
- **Different** types of cement must be stored **separately**
- Bags must be **carried and not dragged** across rough surfaces
- When a bag has been **damaged**, the cement should be transferred as soon as possible into two old cement bags and not a fertiliser or sugar bag
- You must **not walk** across the bags of cement as your shoes/boots may damage or tear the bag. Wear protective footwear, preferably with rubber soles
- **No equipment** must be stacked on top of the bags during transport.

Portable silos are used for the storage of cement **delivered in bulk** to construction sites. The silos, which usually have a capacity of between 12–50 tons, are filled by special bulk transporters operating between the factory and the site.

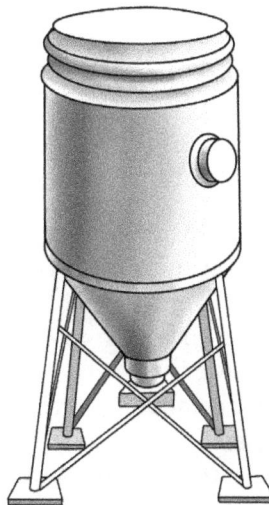

Figure 2.18 A portable silo for storing cement

What do you think are the advantages of cement supplied in bulk rather than in bags?

Cements supplied in **bulk** have certain **advantages** over those in bags:

- Cement in bulk is cheaper
- Storage is compact
- The silo is waterproof
- The cement is automatically used in the order in which it was received
- Wastage due to broken bags and spillage is reduced
- Handling costs are reduced
- Silos usually discharge the cement in controlled quantities into the mixers, reducing under- or overuse.

What precautions do you think need to be taken when using bulk cement?

There are certain **precautions** that need to be adhered to when handling cement in silos:

- Ensure that the silos are clearly marked, indicating the type of cement being stored
- When receiving cement delivered in bulk tankers, check the seals on the tanker to ensure they are not broken
- Check the inspection hatches and filters in weatherproof silos.

When **placing orders**, it is necessary to state at least these items to avoid the wrong material arriving on site:

- Company name
- Site address
- Order number
- Quantity: if ordering large quantities, state whether bagged or bulk cement is required
- Type of material, e.g. CEM III A 32.5. 'Cement' is not descriptive enough.

EXAMPLE 2: Order sheet

Company name			
Site address			
Order no.			
Quantity			
Type of material	Date of order	Date required	Signature

Delivery notes must be inspected. Check that the correct details (as ordered) appear on the delivery note. Then, where possible, check the contents of the delivery vehicle. This will help prevent the wrong material being off-loaded on site. Separating such material and returning it to the supplier is expensive for all concerned. It is easy to check if cement in bags is the right type. Bags should not be accepted if they are wet or broken.

Self-evaluation 2.2

1. Complete the sentences:
   a. Blast-furnace slag is the waste product in the manufacture of _____.
   b. The dominant compounds in cement are _____.
   c. In South Africa the trade name of GGBS is _____.
   d. CSF is a _____, i.e. it reacts with lime in the presence of water.
   e. The abbreviation FA denotes _____.
   f. The cement store should be _____ and solidly constructed.
   g. _____ discharge the cement in controlled quantities.
   h. Arrangement of the packing order should be '_____, _____'.
   i. The material that comes out of the kiln is in the form of dark grey nodules, called _____.

2. State whether the following statements are **true** or **false**:
   a. Cement is made from a mixture of limestone and shale or clay.
   b. The colour of cement plays an important role in defining strength characteristics.
   c. GGBS hydrates more slowly than cement.
   d. Condensed silica fume is not a very fine powder.
   e. CSF blends produce higher strengths than the same amount of ordinary cement.
   f. Stacks should not exceed 20 bags in height.
   g. Silos discharge the cement in the order it was received.
3. Answer the following:
   a. Draw a flow diagram describing the main stages in the manufacture of cement.
   b. What are the raw materials used to make cement?
   c. Describe the important properties of cement in the construction industry.
   d. Distinguish between setting and hardening of cement.
   e. Explain the meaning of 'setting time' for cement.
   f. What is the difference between false set and flash set, and what are they caused by?
   g. Explain the process of hydration of cement.
   h. Why is gypsum added in the manufacture of cement?
   i. What does exothermic mean?
   j. What is a blended cement?
   k. How is the gypsum content in PC specified and why is the amount added carefully controlled?

2.3.2 Water

Water acts as a **lubricant**. For any mix, the higher the water content, the less stiff or sloppier the concrete. The slump (see section 2.2.3) will be higher and the placing and compacting of the concrete will be easier, but segregation is more likely to occur.

It is important to remember that if the water content is increased, but the cement content remains the same, the concrete will be weaker due to a higher w:c ratio.

Water is also needed for the **hydration process** which causes the concrete to set and gain strength. If water is added, it must be clean and not contain any impurities.

Activity 3

This is a simple experiment to check the differences in characteristics when using differing w:c ratios. All you need is cement, water and some mixing bowls.

Mix up these ratios in separate mixing bowls:

- A cup of cement with ½ cup of water
- A cup of cement with 1 cup of water
- A cup of cement with 3 cups of water
- A cup of cement with 6 cups of water.

Make a note of the consistency of each and decide through observation whether the cement strength has changed from mix to mix. Has the strength reduced through the dilution of the cement in the water?

2.3.3 Aggregate

Coarse and **fine aggregate** make up **two-thirds of the volume** of concrete and have important effects on the properties of fresh or hardened concrete. **Stone** is known as **coarse aggregate** and sand is known as **fine aggregate**. A sieve with square openings 4.75 mm in size is used as a 'boundary' between coarse and fine aggregate. Most coarse aggregate particles are retained on this sieve; most fine aggregate particles pass through it. 19 mm stone is commonly used in normal concrete. Aggregates are used in concrete to **reduce the cost** per cubic metre and **reduce shrinkage** and other deformations.

Coarse aggregate (stone)

Coarse aggregate is used in concrete for **bulk** and because it is cheaper than cement, making the mix more **economical**. If the stone size is increased, less water is needed to give the required slump, therefore less cement is necessary to maintain the same water:cement ratio and strength.

Using stone in the mix reduces the paste content and therefore makes the concrete more stable. Larger aggregate has a smaller specific surface area (surface area/unit volume) requiring less water to coat the particles. This also allows for a lower cement content to be used, making the concrete cheaper.

Material from quarry or waste rock dump

Conveyor belt

Primary crusher

Screen

Reject material (recycled)

Secondary crushers

Screens

| 2.36 mm | 4.75 mm | 6.7 mm | 9.5 mm | 13.2 mm | 19.0 mm |

Bins storing various size aggregates (or to ground stockpiles)

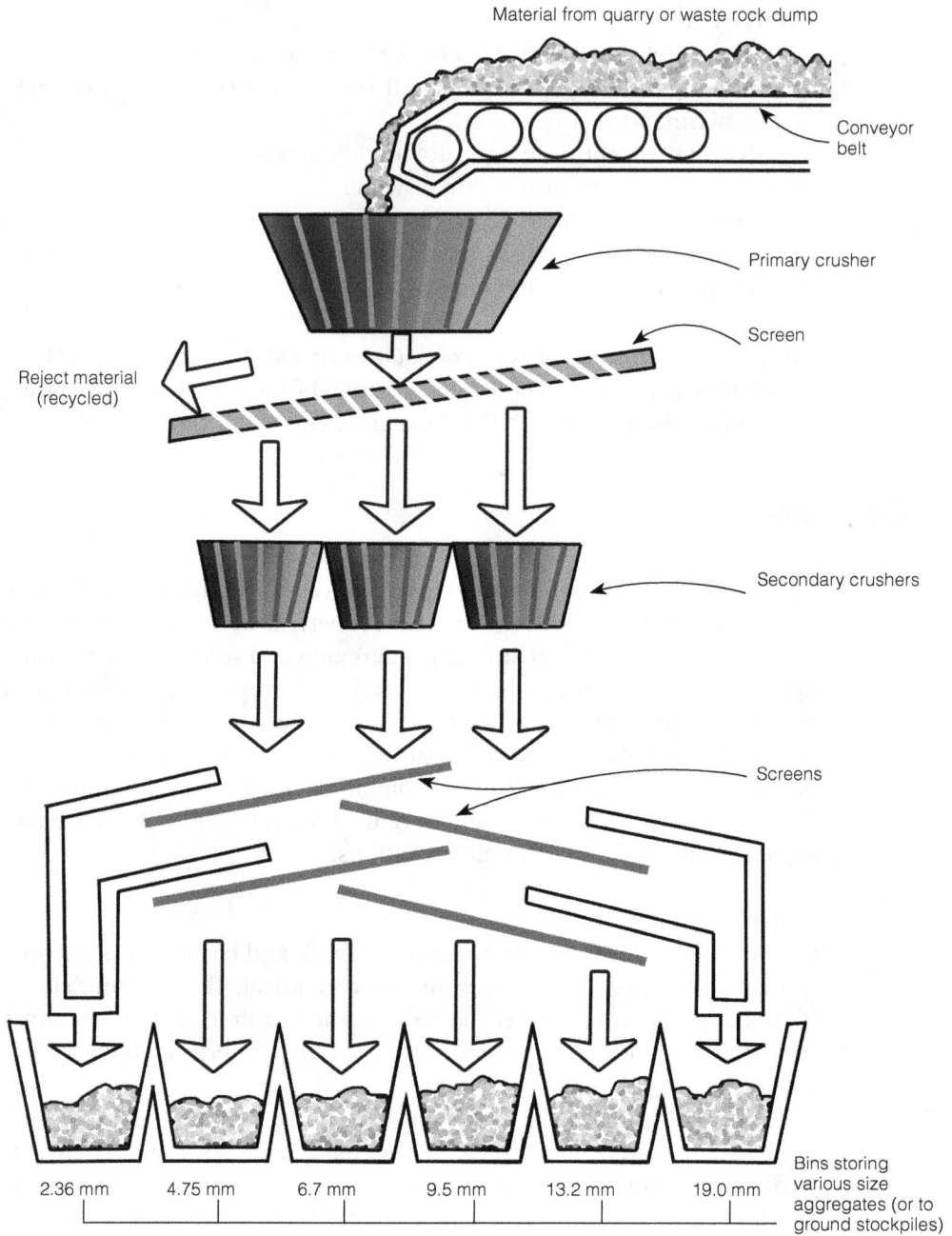

Figure 2.19 A crushing plant at a stone quarry

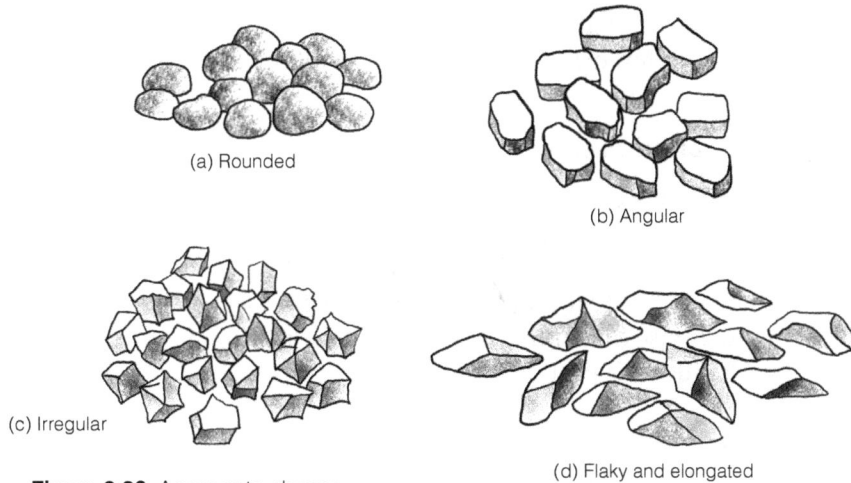

(a) Rounded

(b) Angular

(c) Irregular

(d) Flaky and elongated

Figure 2.20 Aggregate shapes

Most aggregates are derived from natural rock which is blasted and crushed at the rock quarries to produce the various fractions or sizes. The shape of the crushed aggregates ranges from **cubical** to **angular** to **flaky** to **elongated**. The latter two shapes should be avoided in concrete mixes.

Natural aggregates are obtained from river banks, dunes and pits. They are usually better shaped than crushed material. Rounded aggregates make a mix more workable and require less water and less cement. Long, flat, sharp particles make harsh, 'unfriendly' concrete. Many round aggregates have a smooth surface texture. The surface texture refers to the roughness or smoothness of the surface of the aggregate. This reduces the friction in the mix and makes it more workable. Some crushed materials are smooth, but most have a rough surface texture.

Fine aggregate (sand)

Fine aggregates are the aggregates whose size is less than 2 mm, in other words the grains pass through the holes of a 2 mm sieve. Sand is used as coarse aggregate in the preparation of cement and concrete mortar and also as a void filler. Please refer to chapter 1 to refresh your understanding of coarse and fine aggregates – remember that sand is one of the finer fractions of a coarse aggregate.

Fine aggregate is used as a **void filler**. It fills up spaces between the stone and cement. It also affects the amount of water needed in the mix. Badly shaped particles need more water than rounded and smooth particles to give the same slump. Sand also reduces the paste content and makes the concrete more stable. Sands lacking fine fractions – passing through a 300 μm sieve – produce harsh concrete that bleeds and has a tendency to segregate. Ensure that the sand is free from impurities, for example clay, organic matter, etc., at all times.

2.3.4 Chemical admixtures

Admixtures are materials other than water, cement or aggregates that are mixed into concrete to:

- Change its properties in the fresh and/or hardened state
- Compensate for aggregate deficiencies
- Increase the strength gain Delay the setting of concrete
- Reduce the cost of concrete.

Types of admixtures

Admixtures are generally classified according to their functions. The most common are:

- Plasticisers or water-reducing agents (WRA)
- Superplasticisers or high-range water reducers (HRWR)
- Retarders
- Accelerators
- Air-entraining agents (AEA).

With the exception of pigments, most admixtures are in liquid form but some specific types may be available as soluble powder.

Plasticisers or **water-reducing agents (WRA)** increase the slump of concrete at a given water content. Alternatively, they permit a reduction in water for the same slump. Consequently, cement content may be reduced for the same strength.

Superplasticisers or **high-range water reducers (HRWR)** perform the same function as plasticisers but their effect is greater. Superplasticisers form an important ingredient of high-strength concrete.

Retarders delay the setting and hardening of concrete. This may be necessary when placing is delayed or ambient temperatures (atmospheric) are higher than normal.

Accelerators are admixtures that accelerate the **hardening** and/or **early strength development** of concrete. They do not significantly affect the **setting time**. Therefore, accelerators are used when high early strengths are required, for example in repair or marine work.

Air-entraining agents (AEA) entrain (trap) countless very small separate air bubbles in concrete to improve workability, reduce bleeding of the fresh concrete and enhance durability of hardened concrete exposed to cycles of freezing and thawing. Entrained air tends to reduce strength and for that reason mix proportions (water:cement ratio) should be revised accordingly.

Using admixtures

What do you have to keep in mind when using admixtures?

When using admixtures, the following points must be observed:
- The overall effect of the admixture on the fresh and hardened concrete must be assessed by preliminary tests, including trial concrete mixes
- The manufacturer's instructions regarding dosage and time of addition in the mixing cycle must be adhered to
- If material sources change, further trials should be carried out.

The effect of an admixture may vary with the:
- Nature of a particular mix, e.g. proportions
- Type of aggregate used
- Type and source of cementitious material
- Temperature
- Mixing time.

Many cases of overdosing occur, often due to carelessness or the use of incorrectly calibrated dispensers. A **dispenser** is a container that controls the amount of material to be added to a mixture. If it does not do this properly, the result is a mix which is not to specification. Calibration is a means of ensuring that these amounts are correctly controlled.

What happens when overdosing with admixtures occurs?

The effects of overdosing with various admixtures include:
- Retardation of set and/or excessive air-entrainment, when using **water reducers**
- The longer it takes for concrete to set, the longer it takes to gain strength and the longer formwork must be left in place where it could have been used elsewhere

- Rapid set, when using **accelerators** – not allowing enough time for the concrete to be placed before it sets, decreasing workability
- Strength reduction, when using **air-entrainers** – air in the mix decreases density as well as strength
- Retardation of set, greater bleeding and increased segregation, when **superplasticisers** are used.

2.4 Mix proportions and quantities

The materials in concrete, i.e. cement, aggregate, water and admixture (if required), should be proportioned to give the required properties in the fresh and hardened state.

Mixes originating from the United Kingdom, tailored for the **rounded aggregate** particles, are acceptable as 1:2:4 by volume. However, this is not suitable for South African conditions due to the use of **crushed aggregate** which results in rather stony mixes. Mixes in the order of 1:3:3 or 1:4:4, etc. are more suitable to our conditions.

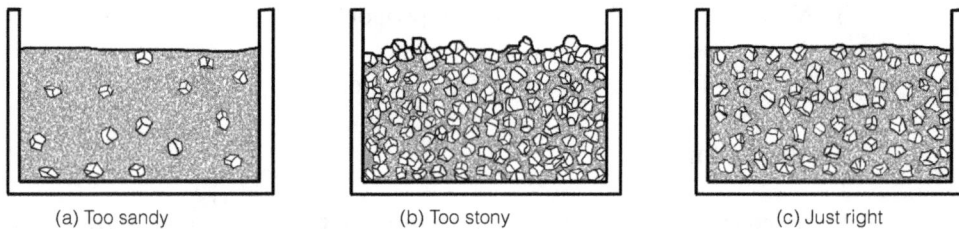

(a) Too sandy   (b) Too stony   (c) Just right

Figure 2.21 A satisfactory sand/stone balance

2.5 Concrete mix design

The objectives of a mix design are to select suitable (economical) materials and to proportion these materials to produce concrete which will satisfy the specific performance requirements (workability, comprehensive strength and durability) as well as give the correct yield or blend, i.e mixture of the right quantities of water, sand, stone and cement for a specified concrete mix.

The mix design is based on the following three rules:

> **Rule1: The volume of the concrete is the sum of the solid volumes of the constituents.**

This implies that the concrete is fully compacted, i.e. it contains no air voids.

$$\text{Solid volume (m}^3\text{)} = \frac{\text{mass (kg)}}{\text{particle relative density} \times 1\,000}$$

For example, one bag of cement contains 50 kg of cement. The **particle relative density (RD)** of CEM 1 may be taken as 3.14, therefore

$$\text{Solid volume of one bag of cement} = \frac{50}{(3.14 \times 1\,000)}$$
$$= 0.0159 \text{ m}^3 \text{ or } 15.9 \text{ litres}$$

> **Rule 2: For any specific materials and conditions of test, the strength of fully compacted concrete depends only on the ratio of water to cement (w:c) in the mix.**

As mentioned earlier in this unit, the relationship between the type of cement used and the ratio of water to cement will have a significant effect on the strength characteristics of the concrete mix.

> **Rule 3: For a given consistency and with given materials, the total amount of water required per unit volume of concrete (the amount of water required) is practically constant regardless of cement content, water:cement ratio, or proportions of aggregate and cement.**

The water requirement of a concrete mix at a given consistency (slump) is determined mainly by two factors: the properties of the sand, i.e. particle shape, surface texture and grading, and the nominal size of the stone.

2.5.1 The design process

What are the five steps in the design process?

Step 1 is selecting a w:c ratio to satisfy the strength requirements.
Step 2 is estimating the water requirement based on the aggregates.

Step 3 is modifying the water requirements to suit the required slump.
Step 4 is calculating the cement content.
Step 5 is determining the stone and sand content of the mix.

Step 1: Selection of w:c ratio

From table 2.6, select the appropriate water:cement ratio to achieve the required characteristic strength. As can be seen, the water to cement ratio is determined by specifying the type of cement used on the horizontal scale to the strength requirement of the mix which is the vertical column. For example, concrete with a required strength of 30 MPa will need a w:c of 0.56, if made with CEM 32.5 cement.

(The table is based on the average performance of South African cements. Strengths of mixes made with specific cements should be verified.)

Table 2.6 Suggested w:c for trial mixes for various characteristic strengths and cements

Characteristic strength of concrete at 28 days, MPa	W:C*				
	CEM strength class 32.5	CEM strength class 42.5	CEM strength class 52.5	70% CEM I 42.5 30% FA	50% CEM I 42.5 50% GGBS
20	0.66	0.72	0.82	0.60	0.64
25	0.61	0.67	0.77	0.55	0.59
30	0.56	0.62	0.72	0.50	0.54
35	0.53	0.57	0.67	0.47	0.51
40	0.50	0.52	0.62	0.44	0.48

* C refers to total cementitious material, e.g. CEM I 42.5 plus FA.

The amount of water to the amount of cement plays an important role in its strength characteristics. The more cement, the stronger the paste, resulting in stronger concrete. The less cement, the more watery the paste will be, resulting in a weaker concrete.

Step 2: Determine the water requirement

Depending on the size of stone used in the mix, the amount of water required per m^3 can be estimated from table 2.7.

Table 2.7 Water requirements of concrete mixes (75 mm slump) for average-quality sand

Nominal size of stone (mm)	Water requirement of concrete (ℓ/m^3)
9.5	235
13.2	225
19.0	(210)
26.5	200
37.5	190

For example, if you are using a 19 mm stone, the water required in the mix would be about 210 ℓ for every cubic metre. This volume of water is based on a 75 mm slump. If the slump requirement is under or above this value then step 3 is required.

Step 3: Modify the water requirement to suit the required slump

Table 2.7 assumes that the slump of the mix is 75 mm. If a slump other than 75 mm is required, then an adjustment must be made using fig 2.22.

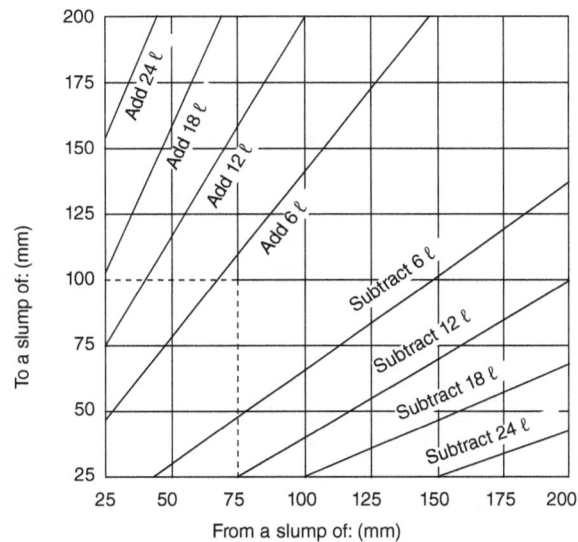

Figure 2.22 Estimating water content adjustment to change slump

For example, the engineer specifies a slump of 100 mm. Starting on the horizontal axis at 75 mm, project a line upwards to strike the line corresponding to the 100 mm slump on the vertical axis. The line does not quite strike the inclined line, so interpolation is necessary. Assume, therefore, that an extra 5 litres of water is required, bringing the amount of water to 210 ℓ +5 ℓ = 215 ℓ.

Step 4: Calculate the cement content

To calculate the cement content, use the following equation:

$$\text{Cement content (kg/m}^3 \text{ of concrete)} = \frac{\text{water requirement in litres}}{\text{water:cement ratio}}$$

Therefore, in the above example,

$$\text{Cement content} = \frac{215}{0.56}$$

$$= 384\,\text{kg}$$

Step 5: Determining the stone and sand content of the mix

The determination of the stone content requires two stages: reading a factor from table 2.8 and then manipulating a formula. The formula for stone content is:

$$\text{St} = \text{CBD}_{\text{St}}(K - 0.1\,\text{FM})$$

where

St	= mass of stone in one cubic metre of concrete (kg/m³)
CBD$_{\text{St}}$	= dry compacted bulk density (CBD) of stone (kg/m³)
K	= a factor that depends on the nominal size of the stone and the workability of the concrete (see table 2.8)
FM	= fineness modulus of sand

Table 2.8 Values for K for determining stone content

Approx. slump range (mm)	Compaction	K			
		Nominal maximum size of stone (mm)			
		9.5	13.2	19.0	26.5
75–150	Hand compaction	0.75	0.84	0.94	1.00
25–100	Moderate vibration	0.80	0.90	1.00	1.06
0–25	Heavy vibration	1.00	1.05	1.05	1.10

Continuing the above example, if you are using 19 mm stone, a sand with a fineness modulus of 2.3, a slump of 100 mm and it is required that the concrete be hand compacted, a K value of 0.94 will be obtained from table 2.8.

Given a CBD of 1 580 kg/m³ for 19 mm stone, the above formula will give:

$$\text{St} = 1\,580(0.94 - 0.1 \times 2.3)$$

$$= 1\,120\,\text{kg/m}^3$$

In order to calculate the sand content per cubic metre, use the method of solid volumes, i.e.:

Volume of sand = 1 – (vol of cement + vol of stone + vol of water)

or

Volume of sand = $\dfrac{\text{mass of sand}}{RD_s}$

$$= 1\,000 - \left[\dfrac{C}{RD_c} + \dfrac{St}{RD_{St}} + \text{volume of H}_2\text{O}\right]$$

where C   = mass of cement per m³ of concrete (kg)
     St   = mass of stone per m³ of concrete (kg)
     RD_s = relative density of sand
     RD_c = relative density of cement
     RD_{st} = relative density of stone

In the above example, if the stone has a relative density of 2.72 then:

Volume of sand = $1\,000 - \left[\dfrac{384}{3.14} + \dfrac{1\,120}{2.72} + 215\right]$

$$= 1\,000 - (122.29 + 411.76 + 215)$$
$$= 250.94$$

But RD of sand is also assumed to be 2.72

$$\therefore\ 250.94 \times 2.72$$
$$= 682.56 \text{ say } 685\text{kg/m}^3$$

The material requirement per m³ of concrete for this example is:
Cement = 384 kg
Stone   = 1 120 kg
Sand   = 685 kg
Water   = 215 ℓ
The mix ratio for this mould therefore is 384:685:1120 or 1:1.8:2.9

Activity 4
By now you would have designed your own home and drawn this electronically (or on paper). Measure the length of all the foundations and using a standard width of 700 mm wide and 250 mm deep, calculate the volume of concrete required (in m³). Once this is done, the following assumptions are to be incorporated:
- Cement type is CEM 32.5
- Particle RD of CEM1 is 3.14
- Strength of foundation concrete required is 20 MPa
- 19 mm stone is to be used in the concrete mix
- RD of 19 mm stone and concrete sand is 2.72, the FM of sand is 2.3
- Concrete slump is 100 mm
- Concrete must be hand compacted.

1. Determine the material requirement for the foundations to your designed home.
2. Contact your local supplier and obtain costs for the various material quantities in order to calculate the overall cost of constructing the foundations.

2.6 Concrete production

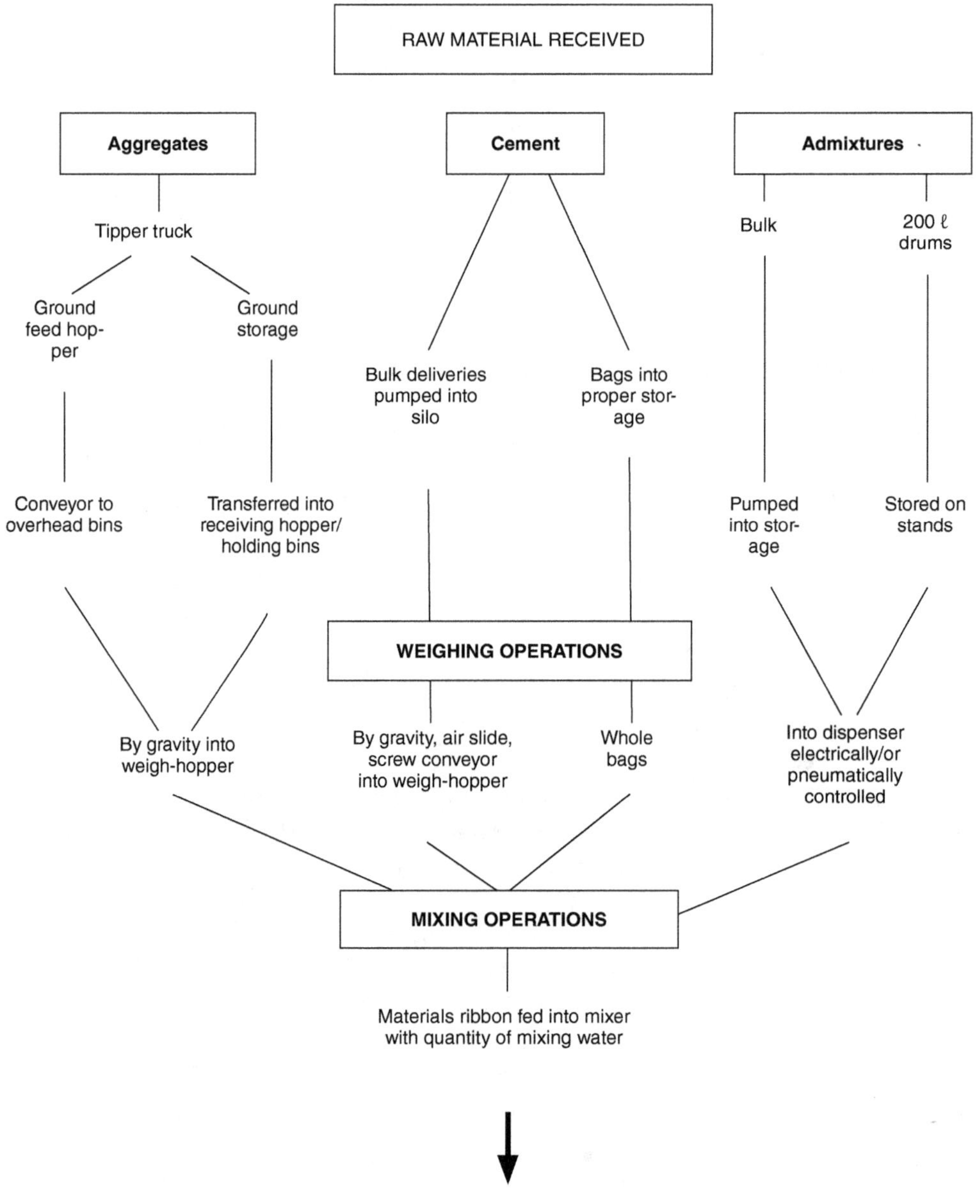

Figure 2.23 Flow diagram of concrete production

Concrete can be produced by either **batching** and **mixing** on site or by purchasing **ready-mixed concrete**. The choice depends on a number of factors:

- Materials properly specified and in sufficient volumes
- Plant (mixer, storage, dumper, front-end loader) either hired or depreciated, and running costs
- Labour
- Whether space is limited on site
- Setting up and maintenance of plant
- Technical expertise
- Quality control, mix designs and calibration.

When using ready-mixed concrete, it is important that there is good communication between **contractor** (purchaser) and **ready-mix supplier** (producer). When ordering concrete, it is advisable to do so well in advance, especially if dealing with large quantities.

There are two ways of specifying ready-mix concrete: **designed mixes** and **prescribed mixes**. Designed mixes are the most common and allow the suppliers to use their expertise to meet the requirements as economically as possible. The onus thus rests on the supplier to design a mix for the contractor's needs.

With prescribed mixes, the contractor (purchaser) specifies the mix proportions and assumes responsibility that the concrete will achieve characteristics of strength, durability, etc.

Table 2.9 Ready-mixed concrete

Designed mixes	Prescribed mixes
Supplier requires the following information:	Contractor specifies the following information:
Strength of concrete	Mix proportions either by mass or volumetric ratio, e.g. 1:3:3
Maximum nominal aggregate size	
Aggregate and nominal size	Cement type
Admixture type (if any)	Aggregate type
Slump at delivery point	Slump or water content at point of delivery
Where the concrete will be used and how it will be transported to workface, i.e. pumping	Admixture type
Procedures to determine if complying with strength requirements.	

2.7 Reinforcement

Reinforcement is the steel rods or bars embedded in the concrete in specified positions. The type and diameter of these bars vary according to the specified requirements.

> *There are generally two types of reinforcement.*

The two types of reinforcement are:
- Mild steel (hot-rolled mild steel plain bars of round cross-section)
- High-yield steel (hot-rolled high-yield stress deformed bars).

Plain round mild steel bars are suitable for most forms of reinforced concrete and can be easily bent, cut and welded. They are denoted with a prefix (R) to indicate a mild steel bar with typical diameters ranging from 8–40 mm.

Bars 16 mm and larger in diameter are generally used for reinforcing foundations, retaining walls and beams. Smaller bars are generally used for floor and roof slabs. In most cases, 8–12 mm bars are used as stirrups and column links. Mild steel bars have a characteristic strength of approximately 230 MPa.

R10

Indicates mild steel   Reflects a 10 mm diameter bar

Figure 2.24 Hot-rolled plain round mild steel bar

High-yield stress deformed bars have a characteristic tensile strength of at least 450 MPa. They can be identified by the ribs running around the bar, compared with the smooth surface of the mild steel bar. The ribs formed on the surface differ from supplier to supplier, and are used to increase the bond between the steel and the concrete. Structures requiring high-strength concrete, for example dams and multi-storey buildings, will usually specify high-yield stress steel. They are available in the same diameters as that of mild steel. High-yield stress bars are denoted with a (Y).

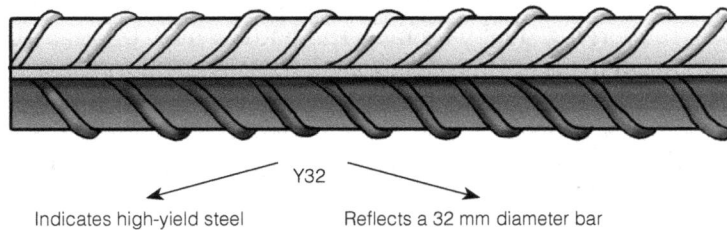

Y32

Indicates high-yield steel   Reflects a 32 mm diameter bar

Figure 2.25 Hot-rolled deformed bars with ribs and high-yield stress

2.7.1 Why concrete is reinforced

Why is concrete reinforced?

Concrete is strong in compression (its ability to be squeezed together) and weak in tension. For this reason reinforcement is used to provide tensile strength to the concrete and in some cases, for example columns, to act as additional compressive strength.

The factors which affect the bond between concrete and steel are:

■ The shape of the reinforcement
■ The reinforcement diameter
■ The compressive strength of the concrete
■ The presence of rust or scale on reinforcement
■ Compaction, bleeding and settlement
■ Cleanliness of the reinforcement, i.e. no oil or dirt.

2.7.2 How reinforcement is used

Reinforcement in concrete is used to control **bending**, **shear** or **compression**.

Bending: Structural members such as beams and slabs are subjected to bending. When a beam or slab, which is supported at its ends, is loaded in the middle, it will tend to sag or bend toward the middle. This causes compression at the top and tension at the bottom. If loading is continued, the beam will start cracking (at the bottom) and will eventually fail (break). Such failure can be prevented by inserting reinforcing steel bars near the bottom face, in order to strengthen the beam to carry more loads.

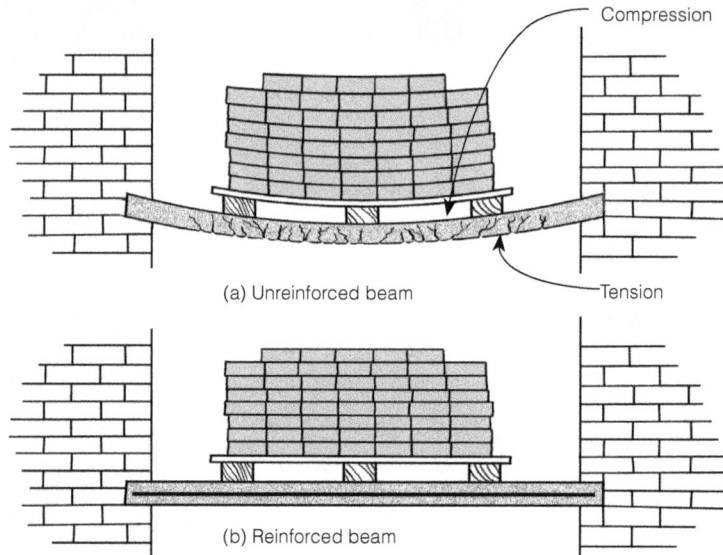

Figure 2.26 Comparison of reinforced and unreinforced beams under load

Shear: A simply supported beam (supported at its ends only) does not only bend, but also has a tendency to move downwards or shear at the supports, resulting in cracks. This is prevented by using inclined bars or vertical stirrups as reinforcement.

Compression: Very tall buildings have high compressive forces transmitted through their structure, especially in the columns. To strengthen these columns and to help carry some of these compressive forces, reinforcement is used. Compression bars are also used on highly stressed, simply supported beams as top steel.

To gain further insight as to how reinforcement is applied in concrete structures, read the *Drawing for Civil Engineering Handbook*.

2.8 Joints

There are various joints used in the construction industry but we will be concentrating on construction joints and movement joints only.

2.8.1 Construction joints

Because it is not often practical to place concrete continuously from the beginning to the end of construction, a structure is subdivided into components of convenient size. These are known as **lifts** or **bays**.

Lift or bay size is influenced by the re-use of formwork, the location of joints for aesthetic purposes and the amount of formwork erection, reinforcement fixing and concreting that can be carried out in a working day.

2.8.2 Movement joints

The two main types of joints are **expansion joints** and **contraction joints**.

Expansion joints are designed to cater for both expansion and contraction of adjacent parts of the structure. They require a sufficiently large gap between the two elements to accommodate the total amount of expected expansion.

Contraction joints are designed to accommodate contraction, usually related to shrinkage of adjacent concrete only. They may accommodate the contraction of a long concrete strip laid in one operation, or the lateral contraction of a number of concrete strips laid together. Contraction joints are spaced more frequently than expansion joints.

2.8.3 Joint fillers and sealants

Where joints are exposed to the exterior environment, they must be **sealed** to prevent ingress of water, dust, etc.

Joint fillers are strips of compressible material such as cork, fibreboard, synthetic rubber or expanded polystyrene. These strips are used to:
- **Form** the joint during construction
- **Fill** the gap during service
- **Support** the joint sealant
- Allow the concrete to **expand** freely.

Fillers must therefore have the following properties:
- **Compressibility** without extrusion (fillers must be able to be compressed/squeezed without jumping/oozing out of the joint, almost like squeezing a tube of toothpaste with the cap on – toothpaste does not come out but it can still move within the tube)
- **Recoverability**
- **Durability**
- **Rigidity** during handling and placing.

Sealing compounds are applied to the groove at the joint face in order to:

■ **Seal** the joint against passage of liquids
■ **Prevent** ingress of grit
■ **Protect** the joint filler.

Sealants are applied in liquid or semi-liquid form and assume the shape of the groove, however they must always be applied acording to the manufacturer's recommendations.

2.9. Precast concrete

Precast concrete is concrete that is cast away from its final position. This could mean that the product is manufactured on site some distance from its intended position, or manufactured at a factory and then transported to the site. The most familiar items are precast concrete garages, culverts, pipes, beams, lintels and fences.

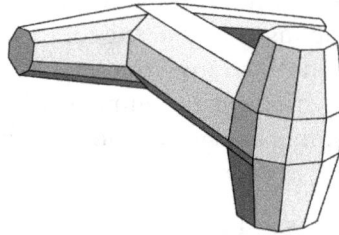

Figure 2.27 Precast concrete dolos

The **advantages** of precasting are:

■ Time saving
■ Economy of moulds or formwork
■ Economy of concrete and reinforcement
■ Ease of producing special features
■ Use of mechanisation in manufacture
■ Reduces space and resources on site
■ Special features and textures can be achieved more easily.

2.10 Prestressed concrete

Fifty years ago prestressing was almost unknown. Since then, however, it has caused the greatest revolution in the art of construction since the advent of mild steel and rolled sections.

Prestressing is a technique of construction whereby initial compression stresses are set up in a structural member to resist or nullify the tensile stresses produced by the load. In other words, prestressing is the creation within a material of a state of stress and deformation that will enable it to better perform its intended function. Most commonly, prestressing creates a compression stress within the concrete that will partially or wholly balance the tensile stresses that will occur while in service.

Activity 5

To illustrate prestressing, lift a number of books from a shelf by putting a hand at each end and then pushing hard. The squeezing force which stops the books from falling is a prestressing force. If necessary, loose books or other items could be added on top and be supported by pressing harder.

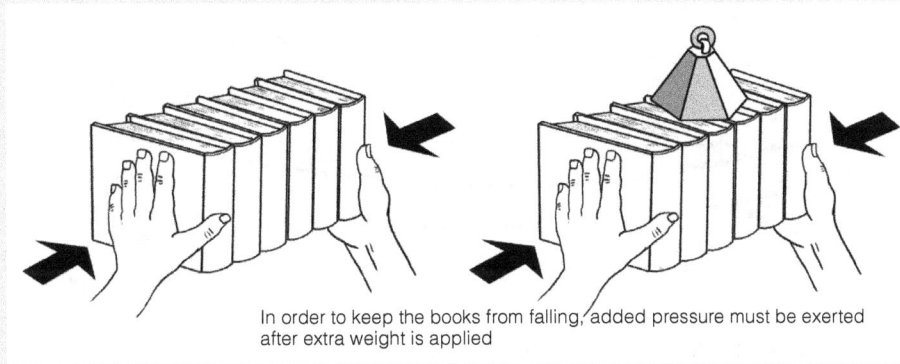

In order to keep the books from falling, added pressure must be exerted after extra weight is applied

Figure 2.28 A prestressing force

Now suppose that instead of books, there are a number of blocks with holes, through which is stretched some form of tendon, for example elastic or spring curtain wire. This tendon would be exerting a force similar to that of your hands. It is now possible to rest this assembly on two supports without it collapsing, even with extra weight added on top.

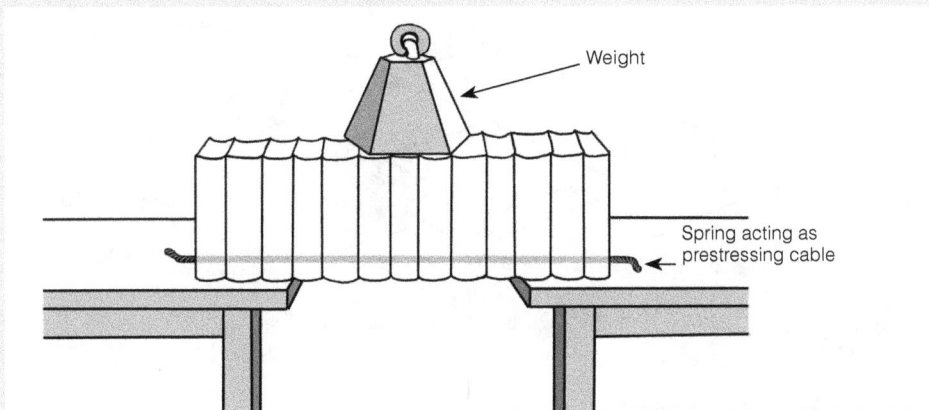

Weight

Spring acting as prestressing cable

Figure 2.29 Books being stressed

There are two ways of prestressing concrete:

- **Pre-tensioning:** The steel is pre-tensioned **prior to casting** and then released once the concrete has hardened
- **Post-tensioning:** Before concrete is poured, a strategically placed metal sheath, or duct, is fixed inside the confines of the member. The concrete is then poured around this pipe which acts as a sleeve or conduit through which the tensioning cables are passed. **Once the concrete has hardened** and gained sufficient strength, the cable can then be pulled or tensioned to a required strength. Once this strength is obtained, the ends of the cable are securely fixed.

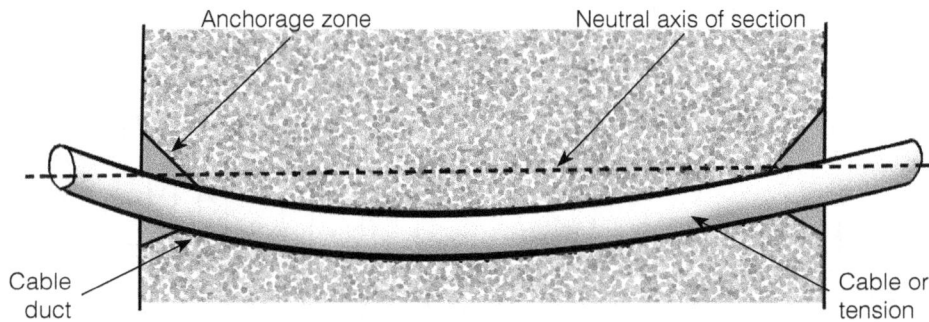

Figure 2.30 A post-stressed beam

2.11 Concrete roads

Although concrete is not a popular choice for constructing roads, mainly due to cost, there are approximately 400 km of concrete roads in South Africa. You will find concrete aprons at airports or toll plazas mainly because of their durability and strength.

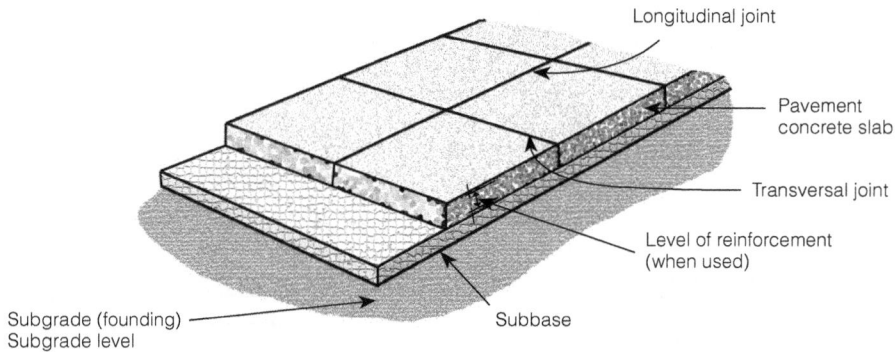

Figure 2.31 A concrete road

Unlike with the normal flexible pavement design, a concrete road structure sometimes does away with a basecourse layer and is constructed on merely a subgrade and often a subbase layer. The road slab may be one of the following:

- Jointed unreinforced slab
- Jointed reinforced slab
- Continuously reinforced slab
- Prestressed slab.

The design thickness of slabs is typically between 100 and 275 mm and is usually designed to a concrete strength of 35 MPa. The **subgrade** can consist of natural ground or imported material that is of uniform density and nonswelling when compacted. The **subbase** is either **granular material** or **cement-treated material** which must be of high quality and compacted to provide the support to the concrete slab. Cement-treated materials are granular materials which have been stabilised by the addition of cement.

Joints in concrete roads are important to control cracking. Therefore, they must be protected by proper sealing to prevent ingress of dirt and water. The most common types of joints used are contraction joints and construction joints.

Self-evaluation 2.3
1. Complete the sentences:
   a. Water acts as a _____.
   b. Particle fractions smaller than 4.75 mm are considered as _____.
   c. Plasticisers increase the _____ of concrete at a given water content.
   d. Accelerators do not affect the _____ of concrete.
   e. There are two ways of specifying ready-mixed concrete, i.e. _____ or _____.
   f. Reinforcement in concrete is used for the following conditions, namely _____ or _____.
   g. _____ are strips of compressible material.
   h. In concrete roads, the joints are to control _____.
2. State whether the following statements are **true** or **false**:
   a. Aggregates are divided into coarse and fine fractions.
   b. Aggregates contribute towards the strength of concrete.
   c. Air-entraining agents improve workability as well as strength of the concrete.
   d. Retarders are used when temperatures are higher than normal.

e.   For prescribed mixes the contractor allows the supplier to design a suitable mix.

f.   A Y20 is a mild steel reinforcing bar, which is 20 mm thick.

g.   Expansion joints cater for both contraction and expansion.

h.   Sealants protect the joints.

i.   Prestressing creates stresses within a member before actual loading stresses occur.

3.   Answer the following:

a.   Why should joints be sealed?

b.   Why are joints provided in concrete?

c.   Why do cooler temperatures retard the strength gain of concrete?

d.   What is meant by the surface texture of aggregate?

e.   Why is the grading of aggregate important with regard to the properties of fresh concrete?

f.   How does the shape of aggregate particles affect the properties of:
  ■ Fresh concrete?
  ■ Hardened concrete?

g.   What is the main difference between the two methods of prestressing?

h.   What are the most common types of steel reinforcement?

2.12 Summary

The purpose of this unit was to:
■ Introduce the basic ingredients of concrete and its characteristics
■ Identify the important factors which influence the properties of concrete
■ Explain and perform some of the tests carried out on fresh and hardened concrete
■ Explain proper storage and handling procedures for cement
■ Describe the special types of cement and their properties
■ Identify the various applications of concrete
■ Explain how concrete is prepared by mixing cement, fine and coarse aggregate, water and possibly admixtures in suitable proportions, to create a blend which is strong and durable
■ Point out that while preparing concrete, various quantities of ingredients should be measured correctly and mixed thoroughly
■ Discuss the importance of prepared plastic concrete having the desired workability suitable for the type of structural component and method of compaction. The cement concrete mix should be designed for the desired strength grade and durability requirements in accordance with the various codes and practices
■ Emphasise the important roles that key aspects, such as ratios and

proper procedures, play in mixing, designing, modifying and laying cement concrete.

Answers

Self-evaluation 2.1

1. a. Paste
   b. Volume, mass, strength
   c. Megapascals (MPa)
   d. Fresh, plastic
   e. Slump test
   f. Segregation
   g. Green concrete
   h. Water:cement ratio
   i. 50, 45
2. a. False, it is influenced by the paste
   b. False, various blends have differing properties
   c. True
   d. False, penetration only to next layer
   e. True
   f. False, it should be too much water, too little fines
   g. True
   h. True
   i. False, will result in weaker concrete
3. a. Separation of the different components (ingredients) that make up the mix will occur resulting in them not binding properly, reducing strength.
   b. We force air out that may be trapped inside the mix.
   c. Concrete consists only of the ingredients used in the mix, i.e. aggregate, sand, cement and water – no steel.
   d. See section 2.3.3.
   e. This ratio determines the strength of the concrete – the higher the ratio, the higher the strength.
   f. In order to achieve full compaction and the highest density attainable, it is necessary that the concrete is workable enough to achieve this.
   g. Consistency could be regarded as the degree of wetness of the mix or sloppiness or stiffness.
   h. It is caused by bleeding.
   i. As cement is expensive, the addition thereof will dramatically increase costs. It is much better to use a material which has similar fine material.
   j. The samples are taken from the back of the truck in carefully divided portions or else from a freshly poured heap on site.

Self-evaluation 2.2

1. a. Pig iron
   b. Calcium silicate
   c. Slagment
   d. Pozzolan
   e. Fly ash
   f. Weatherproof
   g. Silos
   h. 'first in, first out'
   i. Clinker

2. a. True
   b. False, it has no effect
   c. True
   d. False, it is extremely fine
   e. True
   f. False, it should not exceed 12 bags high
   g. True

3. a. You will be required to look outside of this handbook for your answer, but see fig 2.13.
   b. Essentially the raw materials are limestone and shale or clay, but the others as described in section 2.3.1 are also used.
   c. Properties of cement are: it sets and develops strength; it is easily available and cheap; it is versatile; it can be used under water, and it is strong and durable when hard.
   d. Setting is first attained before hardening; the chemical reaction of hydration takes place during setting, whereas the concrete gains strength during the hardening stage; setting is usually of shorter duration, whereas hardening continues for several days.
   e. Setting time describes the process in which the cement stiffens, i.e. it changes from fluid to liquid. This process is when hydration takes place.
   f. False set occurs within a few minutes of mixing with water and no heat is released. Concrete can be remixed without adding water. It is caused by overheating during grinding. Flash set is characterised by the release of heat and is caused by too little gypsum being added to the mix.
   g. When cement and water are mixed, a reaction which releases heat is put in motion. The cement 'grows' fibres that interlock with fibres from nearby cement particles.
   h. It stops the cement setting too quickly.
   i. By dissecting the word into two components, i.e. EXO meaning outside and THERMIC meaning heat, one can ascertain that the word means giving off heat.

j. It is not a pure product but is mixed with other materials.
k. It is controlled through addition of % by weight of cement; overuse could affect the properties of the cement as well as the chemical reactions.

Self-evaluation 2.3
1. a. Lubricant
    b. Fine aggregate
    c. Slump
    d. Setting time
    e. Designed mixes; prescribed mixes
    f. Bending; shear; compression
    g. Joint fillers
    h. Cracking
2. a. True
    b. False, aggregates keep the cost down and give bulk. The paste adds strength
    c. False, although it does improve workability, it reduces strength
    d. True
    e. False, the contractor decides what he wants and takes responsibility
    f. False, the 'Y' stands for high yield and NOT mild steel
    g. True
    h. True
    i. True
3. a. They must be sealed to prevent water and dust getting into them.
    b. To prevent cracking and also as a limit to end off the day's construction.
    c. Cooler temperatures slow the chemical reaction processes (hydration) of concrete.
    d. Surface texture relates to the 'feel' or outside of the aggregate, i.e. the roughness or smoothness of the surface.
    e. Larger aggregates relate to more bulk, less cement and more cost savings.
    f. Rounded aggregates make the mix more workable and require less water. For the hardened concrete, less friction is required to create a compact mix which results in higher densities being obtained.
    g. Prestressing – tensioning is done before the concrete is cast; post-tensioning is after the concrete has hardened.
    h. High yield (Y) and mild (R) reinforcement.

Advanced exercises

1. What volume of concrete is required for the floor of a house which has an area of 250 m^2 and is 150 mm thick?

2. A ready-mixed truck travels at an average speed of 55 km/h. Its capacity is 6 m^3 of concrete and it is delivering to a site 12 km from the mixing plant. It discharges its load and turns around on the site in 30 minutes and takes 10 minutes to reload and turn around at the mixing plant. How much concrete can the truck deliver in an eight-hour day, assuming it starts and finishes at the mixing plant?

3. Describe how changes in the following coarse aggregate properties can affect the workability of a concrete mix:
   a. Maximum aggregate size
   b. Particle shape
   c. Grading.

4. What should the height of each layer in a slump test be?

5. What is the approximate total weight of 1 m^3 of normal concrete when fully compacted?

6. The continuous beam shown below has a cantilever at the right-hand end. As you can see, the outside edge is unsupported. Loads are applied at the positions shown. By drawing a deflected shape, indicate where the tension reinforcement will be required, if:
   a. All three loads are applied together
   b. The load between B and C is then removed.

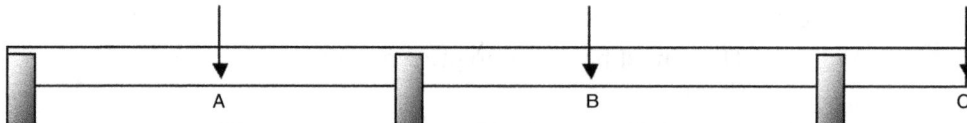

7. What is meant by (a) and (b), what are the harmful effects caused by them and how can they be reduced?
   a. The bleeding of concrete
   b. Plastic shrinkage.

8. Explain how concrete:
   a. Stiffens
   b. Develops strength.

9. List items commonly encountered on construction sites, other than rust and mill scale, that will prevent concrete sticking to steel?

10. What are the main factors that govern the strength of concrete?

11. Why do we cure concrete?

12. What do you understand by the term 'fresh concrete'?

13. What do you understand by the term 'grade of concrete'?

3 Bitumen

Learning outcomes

After studying this unit, you should be able to:

- Describe the production of bitumen by the distillation process
- List the different types of bitumen and discuss the composition and characteristics of each
- Explain the prescribed sampling methods and describe the necessary safety precautions for sampling, and labelling procedures
- Describe and perform the relevant tests that are performed on bitumen, cutback bitumen and bitumen emulsions
- Distinguish between the different applications of bitumen
- Follow the basic health and safety principles for working with bitumen products
- Compare bitumen to tar in their application and properties.

3.1 Introduction

In this unit you will look at the composition, characteristics, treatment and applications of one of the most widely used products in the road industry – **bitumen.** In actual fact, the material that is usually called 'bitumen' is not one specific product, but rather a group of materials called 'bituminous products', which each have slightly different properties. These products include bitumen itself, **asphalt** and **tar.** The type of bituminous product needed for any particular application is manufactured to specific standards.

Figure 3.1 Bitumen being sprayed onto a road surface

Bitumen is the sticky brown or black substance found on the surface of roads. It is also used for roofing and has other applications, but it is especially valuable to engineers as a **binder**. It is useful as a binder because it is strong, adhesive (it sticks to other materials), highly waterproof and durable. It is also resistant to change or damage because it does not react easily with other substances, such as alkalis, acids and solids.

A **binder** is a bitumen- or tar-based material used in road construction to bind together different particles. In other words, a binder acts like a glue.

Bituminous products have been used since ancient times. The Egyptians and the Romans used them in roads, buildings and irrigation systems. Since then an increasing number of applications have been found and nowadays it is used in **roofing, road surfacing, insulating, varnishes** and as a **base for coal tar paints.** However, about 85 % of the bitumen produced today is used as **paving material.**

140

Figure 3.2 A modern highway

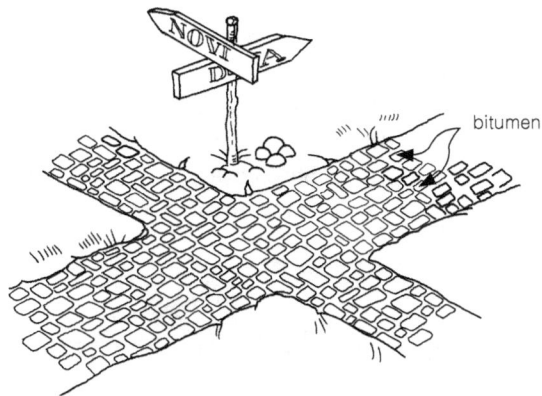

Figure 3.3 An ancient Roman road

3.2 Sources of bitumen

Bitumen occurs naturally in certain places. Two examples of such places are in lakes on the island of Trinidad, where it has been forced through the Earth's crust, and in the pores of certain rocks in Switzerland. However, most of the bitumen used in construction is **synthetically produced.**

Did you know that bitumen is actually a derivative of oil?

Bitumen is extracted from crude oil as the oil passes through the distillation process during the manufacture of petroleum. Refineries such as Shell, Caltex and Sasol manufacture bitumen to strict engineering specifications. A **derivative** is a product that is made from another material. For example, one could say that potato chips are a derivative of a potato.

These refineries do not produce bitumen as their main product. It is generally produced from the materials that remain after the manufacture of their petroleum products – in other words, it is a by-product of the petroleum manufacturing process.

Figure 3.4 An ocean oil rig drilling for crude oil

There are usually differences in the batches of crude oil that are used as a source for bitumen, and so there are similar differences in both the physical and the chemical properties of different batches of bitumen.

3.3 Bitumen and tar

Why do we always talk about tarred roads, even though we are using bitumen in road construction?

It is important to be able to distinguish between tar and bitumen. The term 'tar road' is actually inaccurate. Tar is also used in the road industry but is being phased out because it has a negative effect on the environment. Today, bitumen is generally used in the construction of roads.

Tars do still occasionally occur naturally as tar pits, but are normally derived from the **destructive distillation** of coal. Destructive distillation means to subject a material to heat alone, without access to air. Tar can also be produced from the processing of gas at gasworks. In southern Africa, most of the crude tar pitches (from which various grades of tar are produced) come from two sources:

■ The carbonisation of coal in steel works
■ The synthetic fuel manufacturing process that takes place in, for example, Sasol's refineries; it is a by-product of this process.

Because tars can be degraded by the ultraviolet rays in sunlight, they react more readily with air than bitumen. This explains why tars are generally used in the underlying layers in road works and bitumen for the surfaces.

The differences between road tar and bitumen are set out in table 3.1.

A **solvent** is a liquid that dissolves other substances, for example diesel. Any substance that can cause cancer is called a **carcinogen**. Some materials, when loaded to the elastic limit, break oddly. This is called **brittle** behaviour. Other materials can be loaded beyond their elastic limit into the plastic range without breaking suddenly, but rather deforming gradually – this is called **ductile** behaviour.

Table 3.1 The differences between road tar and bitumen

Tar	Bitumen
Appears **black** when seen in large quantities	Appears **black** when seen in large quantities but **brown** in thin films
Has a strong distinctive smell	The smell is not as distinctive
Is more **brittle** at lower temperatures	It is **less brittle** at low temperatures
Responds easily to **temperature** changes	Does not react as quickly to small temperature fluctuations
Spoils easily when overheated	Is less likely to spoil except in extreme heating conditions
Can be removed from bulk containers more easily than bitumen	Cannot be removed as easily from bulk storage containers
Spilling a **petroleum solvent** onto a tar surface will not affect it adversely	A petroleum solvent spilled on a bitumen surface will **soften** it resulting in a loss of strength
A few of the thousands of chemical constituents of tar are known to be carcinogenic	**Not as carcinogenic** as tar
Tar is **toxic** and must not be inhaled	Is **non-toxic**
Skin and eye irritation is worsened by exposure to the sun's ultraviolet light	In most cases contact with the skin or eyes can be treated with water

Activity 1

Compare the information you are reading with your own experiences and ask yourself questions like these:

1   Does 'carbonisation' simply mean 'burning'?
2   Are synthetic fuel, petroleum, petrol and liquid fuel all the same thing?
3   Try to get some bitumen from a laboratory and then do the following to see how the bitumen reacts:
    - Smell it
    - Heat it
    - Taste it
    - Burn it
    - Stretch it
    - Compress it.

3.4 The production of bitumen

During the apartheid era there was an oil embargo against South Africa – in other words, other countries could not officially sell oil to South Africa. In this time, petroleum or crude oil was obtained from merchants who, in turn, had obtained their oil from the Middle East oil fields. This resulted in a fairly steady supply of crude oil with consistent properties over long periods. As a result, the supplies of bitumen that were produced also had more or less the same properties.

When the oil embargo was abolished, spot market trading was introduced. This has resulted in a situation where crude oil can now be obtained from almost any oil field or supplier. Because the properties of the crude oil supplies now differ, the properties of the various refined bitumen supplies also differ.

Spot market trading takes place when agents trading on behalf of oil companies sell the crude oil to interested buyers while the oil tankers are still at sea.

In South Africa, all bitumen used in road construction is processed at oil refineries in Cape Town, Durban and Sasolburg where imported crude oils are refined to produce petrol, diesel and other petroleum-based products.

Fig 3.5 is a simplified flow chart illustrating the flow of crude oil through a typical refinery. It emphasises that part of the manufacturing process that relates to the production of bitumen.

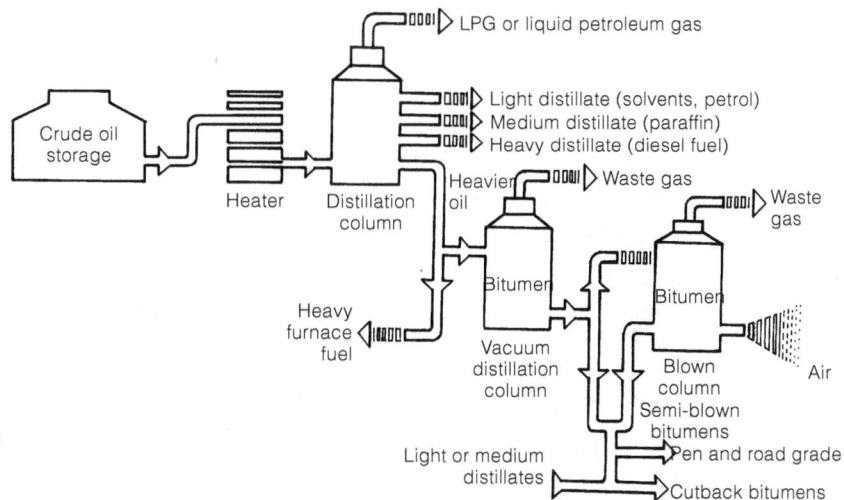

Figure 3.5 The flow of crude oil through a typical refinery, showing the bitumen production process

Crude oil is pumped from large oil tankers into the crude-oil storage facilities. It is then heated and delivered to an **atmospheric distillation column**, a tall steel column in which evaporation at atmospheric pressure takes place. By **evaporation** (when a liquid is heated and changes into a gas) various lighter parts or 'fractions' of the oil are drawn off. These fractions are the less dense vapours of the oil that are created during the boiling process. The vapours that are drawn off include vapours of gas, petrol, paraffin and diesel (see LPG – liquid petroleum gas – in fig 3.5). Some heavy oils and bitumen remain.

These heavy oils and bitumen may be used as fuel, but a large portion of them is processed by further distillation in a vacuum to produce a softer bitumen which is then drawn off separately. Treatment in a vacuum enables oil fractions to be drawn off in vapour form at relatively low temperatures.

To create a harder bitumen, air in the blowing column is forced into the bitumen under carefully controlled temperature conditions. The rate at which the air is blown, its pressure, the temperature and the duration of the process all have an effect on the properties of the final product.

3.5 The composition of bitumen

Bituminous materials are a complex combination of **hydrocarbons**. Bitumen contains small quantities of **sulphur**, **oxygen** and **nitrogen** and trace quantities of metals such as **vanadium**, **nickel**, **iron**, **magnesium** and **calcium**. However, 90–95% of bitumen is made up of hydrocarbons. A result of this is that bitumen can be prone to damage over time: hydrocarbons easily combine with oxygen, and in time this causes changes in bitumen's chemical structure which results in cracks. This is especially true for road surfaces.

The word hydrocarbons can be broken up into hydrogen [H] + carbon [C], which shows that hydrocarbons are molecules made up from a combination of hydrogen and carbon atoms.

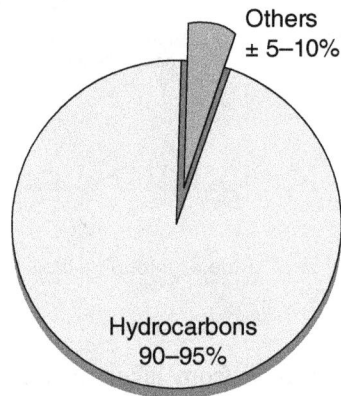

Figure 3.6 The composition of bitumen

Bitumen

Asphaltenes
(insoluble)

Maltenes
(soluble)

Resins   Aromatics   Saturates

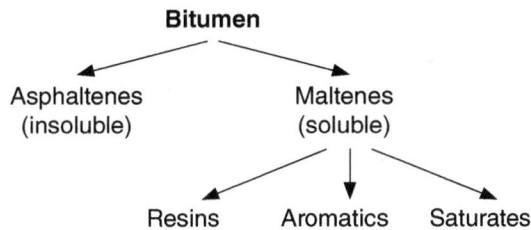

Figure 3.7 Components of bitumen

The chemistry of bitumen is very complicated and, for the purposes of this unit, we will say that it has two different types of components (these components can be separated with the help of chemical techniques). The **soluble** portion is known as **maltenes** (or petrolenes), and the **insoluble**, separate portion as **asphaltenes**. Soluble materials are materials that can dissolve. The maltenes are further subdivided into resins, aromatics and saturates.

Asphaltenes are the hard, black, glassy material constituting 5–25% of the bitumen. Increasing the asphaltene content makes the bitumen harder and more viscous. A viscous material is stiff, thick and resistant to flow.

Maltenes consist of the following:

■ **Resins:** Dark brown solids or semi-solids acting as a 'peptiser', or dispersing (spreading) agent, for the asphaltenes; they affect the rigidity (stiffness) of the bitumen
■ **Aromatics:** A dark brown, viscous liquid making up 40–65% of the bitumen and acting as the major dispersing agent for the peptised asphaltenes
■ **Saturates:** Straw-coloured viscous oils containing both waxy and non-waxy components and making up 5–20% of the bitumen.

It is not yet certain how the quantities of these substances in bitumen affect the performance of the bitumen – in other words, how they affect bitumen's strength and durability.

The molecular structure of bitumen is complex, and the molecules of bitumen samples that have been derived from different sources or that have been blended in different ways vary in size and types of chemical bonding. The three basic molecular structures of bitumen are shown in fig 3.8.

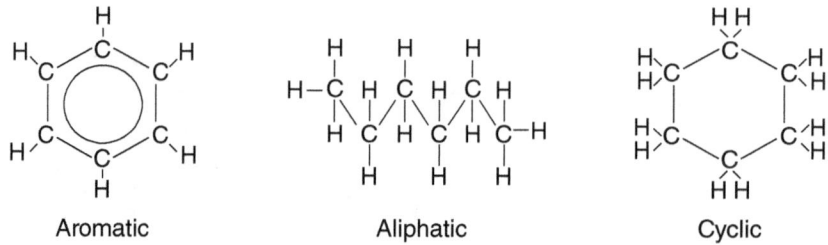

Figure 3.8 The three molecular structures of bitumen

- Aliphatics are linear, three-dimensional chain-like molecules with an oily or waxy composition
- Cyclics are three-dimensional carbon rings with the various atoms attached
- Aromatics comprise flat, stable carbon rings which are stacked together.

Relatively weak chemical bonds hold the molecules together and these can be easily destroyed by heat or shear stress. These weak bonds give bitumen its viscous and elastic characteristics. The molecules in bitumen can be further classified into two functional categories as shown in fig 3.9.

Viscous behaviour indicates that a material deforms with time, for example an oil drop can deform with time. **Creep** is a result of viscous behaviour and is defined as time-dependent deformation under a constant load.

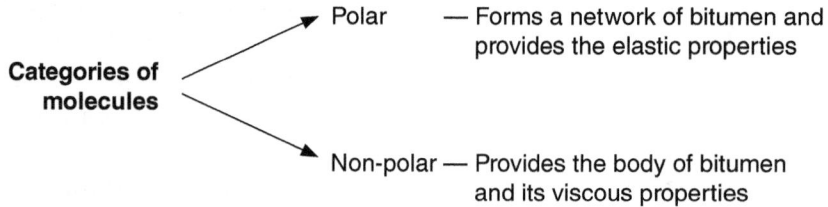

Figure 3.9 The categories of the molecules in bitumen

These two categories of molecules coexist, forming a **homogeneous mixture**. A homogeneous mixture is one that is the same throughout.

Does the behaviour of bitumen change when exposed to heat?

You can try this in a laboratory by heating a cold sample of bitumen – note how its behaviour changes, becoming more fluid. The HIGHER the temperature of the bitumen, the LESS viscous (stiff) it will be – in other words, the hotter it is, the more easily it will flow.

The **balance** between polar and non-polar molecules is important to ensure **good performance**. As bitumen has such a complex and variable chemical and molecular structure, it is extremely difficult to use chemical analysis to determine how the physical properties of a sample of bitumen will influence its performance. Chemical analysis is therefore not very useful as a basis for specifying and selecting bituminous products.

3.6 The behaviour and characteristics of bitumen

Bitumen behaves in five ways:

- **Bitumen is elastic.** When one takes a thread of bitumen from a sample and stretches or elongates it, it has the ability to return to a length close to its original length eventually. For some bitumens this process may take longer than others. This property is referred to as the elastic character of bitumen.

- **Bitumen is plastic.** When temperatures are raised, as well as when a load is applied to bitumen, the bitumen will flow, but will not return to its original position when the load is removed – a condition referred to as plastic behaviour. Applying a load means that you put a weight on the bitumen in order to subject it to stress. This could be in a lab or in the bitumen's final position in the road, and it is done to assess the bitumen's reaction to the load.

- **Bitumen is visco-elastic.** Bitumen has a visco-elastic character: its behaviour may be either viscous or elastic, depending on the temperature or the load it is carrying. At higher temperatures there is more flow or plastic behaviour, while at lower temperatures and short-duration loading, the bitumen tends to be stiff and elastic. At intermediate temperatures it tends to be a combination of the two.

- **Bitumen ages. Ageing** refers to changes in the properties of bitumen over time, which are caused by external conditions. These changes are visible as cracks or crumbling areas. When bitumen is exposed to atmospheric conditions, the bitumen molecules react with oxygen, which results in a change of the structure and composition of the bitumen. This process of combining with oxygen, called **oxidation**, causes the bitumen to become brittle and hard and to change colour from dark brown or black to grey. This change is usually referred to as **oxidative hardening** or **age hardening**. This form of ageing occurs more frequently in warmer climates or during warm seasons, causing older pavements to crack more easily. The condition can also occur where the surface films of bitumen are thin, or if there has been inadequate compaction during construction. To compact materials means to press them together.

■ **Bitumen hardens.** Exposure to ultraviolet (UV) rays and the evaporation of **volatile** components can cause bitumen to harden. A volatile material is a material that can change into a gas very quickly. There are two kinds of hardening: **physical hardening** occurs when waxy crystals form in the bitumen structure, or when asphaltenes agglomerate (clump together). This condition can be reversed if the temperature is raised. **Exudative hardening** is caused by the absorption of oily components in the bitumen.

Because of the way it behaves, bitumen has certain characteristics which makes it one of the most versatile materials available in the civil engineering industry today.

These characteristics can be summed up by these seven simple terms:
■ Sticks – has excellent **adhesive properties**
■ Stops water – is **waterproof**
■ Stretches – is **ductile and flexible**
■ Softens – when heated, but **returns to its original** solid or viscous state on cooling
■ Soluble – by **petroleum solvents**
■ Safe – **non-toxic** and very suitable for use in potable (drinkable) water systems
■ Stable – **durable**, with a long life if handled correctly.

Have you noticed the condition of the surface where there are diesel spillages? What about a premix at service stations where spillages sometimes happen?

Petrol is not one of the petroleum solvents that can dissolve bitumen, because it evaporates when it is spilled on the surface, Diesel is a real danger, because it softens the bitumen.

Remember the softening characteristic when you see bitumen in a solid or semi-solid state at ordinary temperatures. As a solid it is in a viscous state, or a state of high viscosity. However, when it is heated it becomes more like a liquid, a condition which we call a state of low viscosity. When it cools down again, it returns to its original solid state.

Do you now understand what happens to bitumen when the temperature changes?

So you're saying a higher temperature equals a lower viscosity and vice versa.

Activity 2

Form into groups of about four to discuss and answers to these questions. Compare your answers with the other groups.

1. Describe how crude oil for bitumen manufacture is obtained, and how bitumen is produced from it during the distillation process.
2. In the form of a table, list the differences between bitumen and tar with regard to their origin, manufacture and uses. You must also mention the environmental consequences of using tar, as opposed to bitumen, especially now that environmental concerns are receiving priority.
3. Explain the chemical composition of bitumen, which essentially consists of two elements, namely carbon and hydrogen and their chains.
4. List the various components of bitumen and discuss their significance.
5. Describe how bitumen behaves when it is exposed to temperature, atmospheric conditions or loading.
6. What is a binder? Is there a difference between bitumen and a binder and if so, what is it?
7. Define the following terms:
   - Hydrocarbons
   - Viscosity
   - Visco-elasticity
   - Tar.
8. Describe the three basic molecular structures of bitumen and their significance.

3.7 Types of bitumen

Several types of bitumen are being marketed in South Africa today. Most of these have been manufactured in accordance with the criteria for standard penetration bitumen (explained in section 3.7.1). Various grades of bitumen are formed by blending hard and soft bitumen (that is, bitumen with a high and a low viscosity) to create bitumens of intermediate viscosity. The various types of bitumen are classified according to their consistency, as measured by either the **penetration** that can be applied to them, or by their **viscosity** and **softening point**. The consistency of a substance is its degree of smoothness and thickness (viscosity). Table 3.2 shows which classifying property or properties are applied to each type of bitumen.

Table 3.2 Classifying properties of bitumen

Type of bitumen	Classifying property
1. Penetration grade (road grade) bitumens	Penetration
2. Cutback bitumens	Viscosity
3. Blown grade bitumens	Softening point and penetration
4. Bitumen emulsions	Viscosity
5. Polymer-modified binders	Viscosity

3.7.1 Penetration grade bitumens

As we mentioned earlier, some bitumens are obtained by fractioning in a distillation column. **Fractioning** is the extracting of vapours of different density during the heating of oil. These bitumens are classified according to hardness. The **hardness** of a sample is determined by inserting a specific type of needle into it, and measuring the depth to which the needle is able to penetrate the sample within a given time and temperature. This test is called the **standard penetration test**, and the bitumen that is tested in this way is called penetration grade or road grade bitumen (see section 3.9.1).

Because different samples of penetration grade bitumen have different chemical compositions and their manufacturing is controlled in different ways, they may vary in consistency from semi-solid at room temperature to semi-liquid under the same conditions.

The penetration numbers obtained by means of the penetration test indicate the hardness of the bitumen. Fifteen 'pen' bitumen is the hardest and 450 'pen' bitumen, the softest. In the construction industry, the term 'penetration grade bitumen' is often abbreviated to 'pen bitumen' or simply 'pen bit'.

Pen bit 80/100

Minimum
penetration

Maximum
penetration

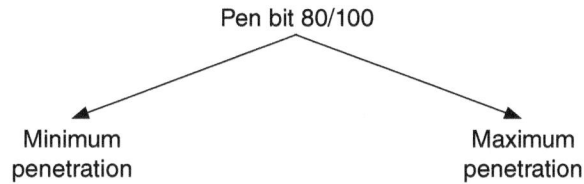

Figure 3.10 The hardness of bitumen is measured by penetration

Penetration grade bitumen forms the basis of other road binders such as cutbacks, emulsions and polymer-modified binders, which are all discussed below.

3.7.2 Cutback bitumens

Cutback bitumens are a range of binders that are produced by **blending** (mixing) penetration grade bitumen and a hydrocarbon solvent, such as paraffin or mineral turpentine.

When the solvent has evaporated, the binder returns to its original penetration grade to tie the particles together. Cutback bitumen gets its name from the solvent that is involved in the process, because the solvent '**cuts back**' or **evaporates**, leaving behind the binder to 'get on with the job'. The solvent used in cutback bitumen is called the 'cutter' or 'flux'.

Three types of solvents are used for the blending process: slow-curing, medium-curing or rapid-curing solvents. The latter two are the most common in South Africa. The choice of solvent determines the rate at which the bitumen will cure when exposed to air. A **rapid-curing** (RC) solvent will evaporate faster than a **medium-curing** (MC) solvent. Curing relates to the evaporation rate of the solvent which influences the setting time of the bitumen. The viscosity of the cutback bitumen is determined by the proportion of solvent added: the higher the proportion of solvent, the lower the viscosity of the cutback.

Cutbacks differ from penetration grade bitumen in that they are more workable – in other words, they can be more easily reshaped. Less heat is required to liquefy cutback bitumen than penetration bitumen, making it easier to use at lower temperatures.

Typical cutback bitumens are MC 30 and RC 250. The letters in the name refer to the curing action of the solvent, and the number to the viscosity of the binder.

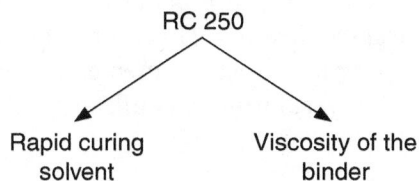

RC 250

Rapid curing
solvent

Viscosity of the
binder

Figure 3.11 A typical cutback bitumen

3.7.3 Blown grade bitumens

Do you remember that we mentioned air is blown through the bitumen to make it harder?

Turn back to fig 3.5 and look at the section showing the blowing column. This is how we obtain blown grade bitumens. These bitumens are more rubbery than penetration grade bitumens, behave better under cold temperature conditions and have a higher softening point. (Softening point is discussed further in section 3.9.1).

3.7.4 Bitumen emulsions

An emulsion is a stable mixture of extremely small particles of one liquid dispersed (spread) in another liquid in which it does not dissolve. A bitumen emulsion, for example, consists of fine droplets of bitumen dispersed in a liquid such as water, in which it cannot dissolve.

Can you think of an example where this occurs? What about when I mix cooking oil and water?

An emulsion can be created by using an **emulsifier** or **emulsifying agent** which is a soap-like substance used to coat each individual particle of another substance and so prevent these particles from sticking together.

Bitumen emulsions are usually manufactured by making bitumen droplets disperse in water. However, it can also be done the other way round, creating an invert bitumen emulsion: water can be dispersed in the bitumen. Invert bitumen emulsions are generally used in cold mixes or for stabilising soil.

Most of the important properties of an emulsion depend on the type and amount of emulsifier used. Emulsifiers are divided into two categories: **anionic** and **cationic** emulsifiers.

Anionic emulsifiers: Anionic emulsifiers are **alkaline** substances which, when they are dissolved, become soluble in bitumen and attach themselves to the bitumen droplets in such a way that each particle is coated with a **negatively charged** film. As a result, the particles repel each other (push each other away) when they come into contact. An anionic emulsion is therefore a negatively charged bitumen-and-water solution.

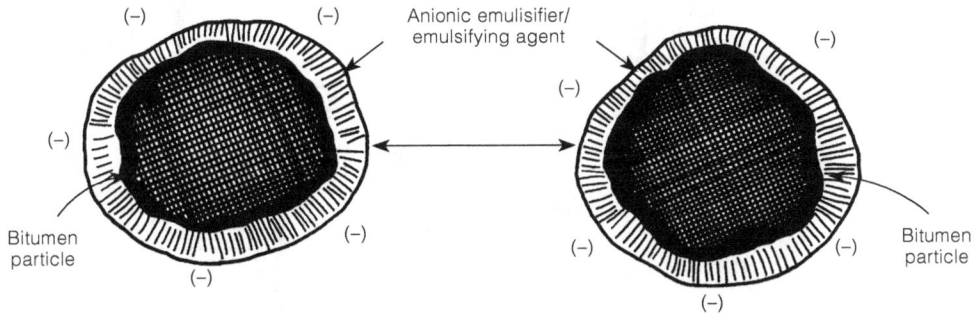

Figure 3.12 A bitumen particle coated with a negatively charged film

Cationic emulsifiers: Cationic or **acidic** emulsifiers are the most commonly used type of emulsifier in southern Africa. Unlike the anionic emulsifiers, these substances release particles with a positive charge which attach themselves to the bitumen droplets, coating them with a **positively charged** film. The result is the same as with the anionic emulsifiers: the particles repel one another and so are prevented from sticking together.

The reason why this type is so popular is that most of the aggregates used in road construction are negatively charged or neutral. This then causes an instant attraction as positive (+) and negative (−) attract.

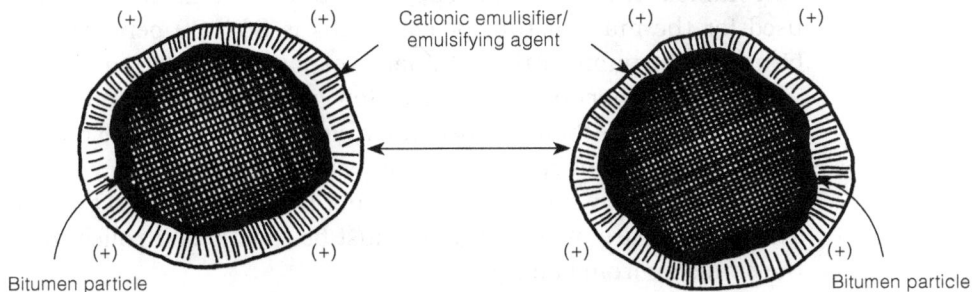

Figure 3.13 A bitumen particle coated with a positively charged film

The manufacture of bitumen emulsions

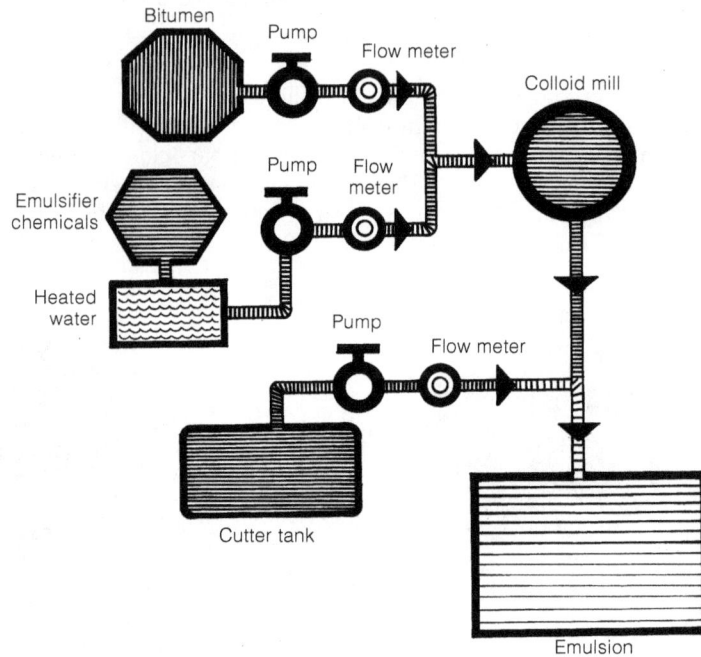

Figure 3.14 Manufacture of emulsions

The manufacture of an emulsion begins with the so-called **water phase**. Emulsion manufacturers use different water phase systems to produce anionic and cationic emulsions. The emulsifier chemicals are dissolved in heated water and are **saponified** with an **alkaline soap**, in the case of anionic emulsions, or with an **acidified soap**, in the case of cationic emulsions. **Saponified** means the individual bitumen droplet is coated or surrounded with a layer of soapy substance. The normal base bitumen used for the manufacture of emulsions is 80/100 penetration grade bitumen (the name of this bitumen is explained in section 3.7.1).

Controlled portions of penetration grade bitumen and chemically treated water are then pumped into a colloid mill. A colloid mill consists of a conical disc or rotor which spins around a stationary part, called the stator, at very high speed, breaking the bitumen into fine droplets. The temperature of the bitumen must be strictly controlled to prevent the emulsion from boiling.

On discharge from the mill, the newly formed emulsion is allowed to cool in storage tanks, each of which contains a particular type and grade of emulsion.

On site or in civil engineering circles, you will often hear people mention 'breaking' when they are talking about emulsion applications. This 'breaking' refers to the evaporation of the water part of the emulsion after it has been applied to the road surface. Once the water has evaporated, the bitumen particles rejoin to form a continuous film which covers the aggregate. There are various factors which can cause this process:

- Loss of water through evaporation
- Loss of water when the water is absorbed by the aggregate
- Chemical coagulation (thickening) of the emulsion, caused by a reaction between the aggregate and the emulsion
- Mechanical disturbance, such as mixing, rolling or traffic action
- The chemical composition of the emulsion.

As a rule, anionic emulsions 'break' because of environmental conditions, whereas cationic emulsions 'break' because of chemical reactions.

Advantages of bitumen emulsions

Bitumen emulsions have the following advantages:

- Emulsions can be used at lower application temperatures than other road grade bitumens, and can be sprayed at lower application rates
- Emulsions are used extensively in slurries, a mixture of fine aggregate, cement and bitumen emulsions used to fill the holes in the final layer of a road surface
- Emulsions can be used on damp or slightly dusty aggregates
- Emulsions save energy because when they are used in a mix, the mix does not need to be heated to very high temperatures
- Emulsions also conserve energy because a petroleum solvent is not required to make them liquid
- Emulsions are environmentally safe, because solvents, which evaporate into the atmosphere, are not required in their use
- They are safer to use than hot binders or than other bituminous products, which in some cases have carcinogenic (cancer-causing) properties
- In South Africa we often use alkaline aggregates, for example **granite** and **quartzite**, which are negatively charged. These aggregates are well suited to the use of positively charged cationic emulsions, which provide good adhesion
- Anionic (alkaline) emulsions are very suitable for use with positively charged aggregates such as **dolomite** and **limestone**
- Cationic emulsions have a built-in anti-stripping agent. This is an advantage if rain falls soon after the emulsion has been applied, because it prevents the rain from loosening the stone chippings.

3.7.5 Polymer-modified binders

Bitumen products that are exposed to **rough** or **severe conditions** such as steep gradients, a high traffic loading, or high road surface temperatures need to perform better than ordinary bitumen products. To improve the properties of bitumen, we add polymers such as natural or synthetic fibres or PVC (polyvinyl chloride) to the bitumen to form 'modified binders'. Polymers are substances made of many small molecules bonded together in a repeating sequence. Modified binders perform better with regard to:

- Adhesion
- Durability
- Elasticity
- Deformation resistance (to deform means to change shape. Because bitumen will readily take on the shape that it is placed in, we add the polymer to strengthen it so that it will resist changing shape)
- Flexibility at low temperatures
- High viscosity at high temperatures.

Three types of polymer-modified binders are frequently used:

- **Styrene-butadiene-rubber (SBR) binders**: These binders are available in the form of an anionic emulsion, which makes blending with emulsified bitumen easier. SBR-modified bitumen is extensively used as cold-applied bitumen emulsions, as well as hot-applied binders, in places like lightly cracked surfaces
- **Styrene-butadiene-styrene (SBS) binders**: These binders exhibit good toughness and elastic recovery even at low temperatures. Hot-applied SBS-modified binders are used for resealing cracked roads
- **Ethylene-vinyl-acetate (EVA) binders**: These polymers can be used to make asphalt mixes more workable and to improve their resistance to permanent deformation. It is therefore an effective substance to use at intersections and on steep gradients. In southern Africa, however, this type of modified binder has only been used on a limited scale up to now.

Non-homogeneous and homogeneous binders

Broadly speaking, there are two types of polymer-modified binders: non-homogeneous and homogeneous binders. (Remember, a homogeneous substance is a substance that is the same throughout.) Non-homogeneous binders are produced by adding **rubber crumbs** to hot bitumen. The binder performs best when the rubber crumbs have only partially dissolved in the bitumen. The presence of undissolved crumb particles gives the binder its non-homogeneous nature. This type of polymer-modified bitumen is usually called **bitumen rubber**. It is the most widely used modified binder in southern Africa. The two substances mainly used to produce bitumen rubber are rubber crumbs and latex.

Rubber crumbs are produced when old or scrap tyres are recycled by grinding them down, in large mills, to a particle size of less than 1.18 mm. About 20% of the rubber crumbs are blended with bitumen at a mixing temperature of between 170 and 210 °C for a period of one to four hours to allow for digestion of the rubber.

Latex is the natural product of the rubber tree and is a suspension of rubber droplets in a watery mixture. Latex combines very well with bitumen emulsions, but due to the presence of water, foaming may occur if it is added to a binder which is heated above 100 °C.

The second group of polymer-modified binders is the homogeneous binders. As the name suggests, the polymers that are added to and blended with the binder completely disperse within the binder. Homogeneous binders, therefore, are binders which contain no solid or semi-solid particles.

Homogeneous binders are more stable than non-homogeneous binders over longer periods of storage at high temperatures. These products may be cooled in storage and reheated when required, as long as they are not subjected to excessive heat.

Activity 3

You have probably noticed that the above content lends itself to the use of mind maps. Create different mind maps of the main sections.

Self-evaluation 3.1

1. Complete the sentences:
   a. Various grades of bitumen are formed by blending _____ and _____ bitumen to create those of _____ viscosity.
   b. The various types of bitumen are classified according to their _____ as measured by either the _____ that can be applied to them, or by their _____ and _____ _____.
   c. The solvent used in cutback bitumen is also referred to as the '_____' or '_____'.
   d. Treatment of bitumen by air blowing results in somewhat more _____ products.
   e. An _____ bitumen emulsion is created when the water is dispersed in the bitumen.
   f. Anionic bitumen emulsions are _____ charged bitumen droplets in water.

g. Breaking of an emulsion refers to the _____ of the water after application.

2. State whether the following statements are **true** or **false**:
    a. Penetration grade bitumen is classified according to its viscosity.
    b. Penetration grade bitumen forms the basis of other road binders.
    c. In the manufacture of cutback bitumen, two types of solvent are used.
    d. The curing action relates to the letters RC, MC or SC that form part of the names of cutback bitumen.
    e. An emulsifier prevents bitumen droplets from sticking together.
    f. The manufacture of bitumen emulsion begins with the heating of the bitumen.
    g. The addition of polymers to bitumen creates modified binders.

3. Answer the following:
    a. Identify the standard types of bitumen, i.e. penetration grade (road grade) bitumen, cutbacks, emulsions, blown grade and polymer-modified binders.
    b. Explain the manufacture of bitumen emulsions with the help of a simplified diagram.
    c. Discuss the application of the various types of emulsions and compare the uses of emulsions to those of ordinary bitumen.
    d. Explain what modified binders are, and identify the different types of modified binders and their uses.
    e. What is meant by a homogeneous modified binder?
    f. What distinguishes a homogeneous binder from a non-homogeneous binder?
    g. What types of modified binders are available in South Africa today?

3.8 Sampling

In using bitumen, careful and accurate sampling is extremely important to ensure that the product is of a high quality and suitable for its purpose. **Sampling** means to select and test small amounts of a product. In South Africa, methods for sampling of road construction materials are prescribed in official documents such as those in table 3.3. Procedures detailed in TMH5 or ASTM D140 documents are recommended when it is necessary to sample either liquid (hot) or solid (cold) binders.

Table 3.3 Official documents for sampling of road construction materials

Sampling of bitumens	Sampling of tars
Technical methods for highways manual 5 (TMH5) issued by the Department of Transport	Standardisation of tar products test committee (STPTC) method GP3
SANS specification in accordance with American Society for Testing and Materials (ASTM) Method D140	Specifications based on the above

3.8.1 Size of binder samples

The appropriate sizes of binder samples are shown in table 3.4.

Table 3.4 Sizes of binder samples

From drums or cartons	1–1.5 kg
From bulk storage tanks and bulk vehicles	4 litres
From drums	1 litre

3.8.2 Containers

The containers used to carry samples should be of a size appropriate to the sample quantities. For liquid binders (hot bitumens, tars, cutbacks and emulsions) use wide-mouthed cans with tight-fitting screw-top or friction-fitting lids. For solid cold binders, use heavy-gauge plastic bags. All containers should be absolutely clean and free from **contaminants** (something that makes a substance impure).

Figure 3.15 A wide-mouthed can with a screw-top lid and a heavy-gauge plastic bag suitable for carrying samples of binders

3.8.3 Sampling method

Samples should be taken from tanks or bulk vehicles by means of a **thief sampler** or through a **sampling valve**. A thief sampler is a device in the form of a can which is lowered to the centre of a load and then filled with binder. A sampling valve is a valve that is built into the storage tank and maintained at the same temperature as the contents of the tank. It may also be a valve that has been installed in the discharge line between the piping on the vehicle (for unloading) and the hose that carries the product to or from the storage tank.

Samples taken from a vehicle that is being unloaded should be drawn when at least one third, and not more than two thirds of the product, has been unloaded or loaded.

When sampling, the following steps should be followed:

- **Step 1:** If all the containers have the same batch number, one drum should be sampled and tested against specifications.
- **Step 2:** If it is found not to comply with specification, then a certain number of other containers must also be sampled. This number is the **cube root** of the total number of containers in the lot. This also applies if a shipment is not all from a single batch. The cube root of a number is an amount that equals this number if it is multiplied by itself twice. For example, if 125 drums were delivered to site, the cube root of these drums would be 5, because $125 = 5 \times 5 \times 5$.
- **Step 3:** In the case of cutbacks, emulsions and polymer-modified binders, the drum contents should be thoroughly mixed by placing the drum on its side and rolling it to and fro, and the sample should be taken with a thief sampler.
- **Step 4:** Samples of solid binders (e.g. cold, hard bitumens) should be chopped out or, for softer grades, cut out with a stiff putty knife. Samples should be taken at least 75 mm from the sides or ends of the container.

When many containers must be sampled, not less than 0.1 kg should be taken from each container, until a composite sample of about 4 litres from each batch has been accumulated for testing. Remember not to take all 4 litres from the same drum as this would not be a representative sample.

3.8.4 Labelling

- Immediately after the sample has been packed, the container should be clearly labelled with the following information: Date
- Product

- Batch number
- Source
- Supplier's delivery note number with labels
- Complaint, or properties to be determined.

Figure 3.16 A labelled container

Why do you think labels must be securely fixed to the containers, not the lid?

If labels have been placed on lids, and many lids are lying around after the containers have been opened, it is very easy to become confused about which lid belongs to which container. The wrong lid may then very well be placed on a container. This problem can be avoided by labelling the containers themselves.

3.8.5 Precautions

Normal safety precautions must be taken when handling hot binders and binders containing volatile solvents. The following basic precautions should be observed:

- Protective clothing should be worn, leaving no exposed skin
- No smoking should be permitted, particularly when handling cutback bitumens, since the solvents used in cutbacks are highly flammable
- While samples are being drawn, the containers must not be held in the hands
- Tongs, or some other device, should be used
- Samples should be drawn slowly, to reduce the danger of splashing the hot binder

- The person taking the sample should stand on the windward side (the direction the wind is blowing from) of the sampling valve and preferably at a level higher than the container
- When closing the filled container, it should be placed on a firm, level surface to prevent accidental spillage of the sample.

Figure 3.17 Safety precautions when handling hot binders

Activity 4
You should now be able to answer questions such as the following:
1. Explain the importance of sampling and the techniques involved.
2. Discuss basic precautions when dealing with bituminous products.
3. Evaluate the differences between the procedures for samples taken from bulk supply or those obtained from drums.
4. In your own words, explain the necessity of sampling bitumen supplied to site.
5. Why is it necessary to establish technical manuals and procedures for both the sampling as well as the testing of bitumen?
6. By what means are bitumen samples obtained from bulk containers?
7. Why is it important to ensure that your bitumen samples are not contaminated in any way?
8. Why do we have to label the samples?
9. What information must be reflected on our labels?

10. A total of 64 drums of bitumen were delivered to site, all bearing the same batch number. When a sample was taken from one drum and tested by the laboratory, it was found not to be according to specifications. How many drums does the technician have to set aside for sampling?

11. When sampling bitumen, certain health, safety and environmental factors must be observed. What are these and why are they necessary?

3.9 Tests on bitumen

The most careful specifications with regard to the design and construction of a bituminous road surface are of little value if the properties of the bituminous binder used in the design are not adequately controlled. To help the engineer or technician ensure that the material has the desired qualities, a number of tests have been devised which measure various binder properties for particular reasons.

In southern Africa, various methods for testing bitumens and cutbacks are followed. These methods are based on those prescribed by organisations such as the American Society for Testing and Materials (ASTM), American Association of State Highway and Transportation Officials (AASHTO) and the Deutsches Institut für Normung (DIN).

Emulsion test methods are laid down in ASTM or are included as part of the SANS 309 specifications (for anionic emulsions) and the SANS 548 specifications (for cationic emulsions).

The test methods applicable to tars are laid down by the Standardisation of Tar Products Test Committee (STPTC) in its publication *Standard Test Methods for Testing Tar and its Products*.

Tests on bitumens can be classified according to **consistency** or **composition**, as follows:

■ **Consistency** is a function of viscosity and refers to the resistance of the bitumen to flow. There are a number of different tests that can be carried out to determine the consistency, each of which has certain advantages under specific conditions

■ **Composition** tests are carried out to determine if the sample contains components in the correct proportions, that is the proportions that were specified in the specifications for its manufacture.

3.9.1 Tests on penetration grade bitumens

Penetration test

In a penetration test, which we briefly referred to in section 3.7.1, it is established how far a standard needle under standard conditions will vertically penetrate a sample of bitumen.

The penetration is defined as the distance, in **tenths of a millimetre**, that this standard needle will penetrate into the bitumen under a load of 100 grams applied for 5 seconds while the bitumen is maintained at a constant temperature of 25 °C.

The test measures the **consistency** or **relative hardness** of penetration grade bitumens, and was formally used as a basis for classifying these bitumens into standard grades in accordance with SANS 307. This test, however, does **not determine** the **quality** of the bitumen, only hardness.

The values that are obtained in the tests are expressed by two numbers separated by a slash, for example '80/100 pen bitumen'. The numbers are the distance, in millimetres, that the needle can penetrate into the bitumen. The first number is the minimum penetration, and the second number the maximum.

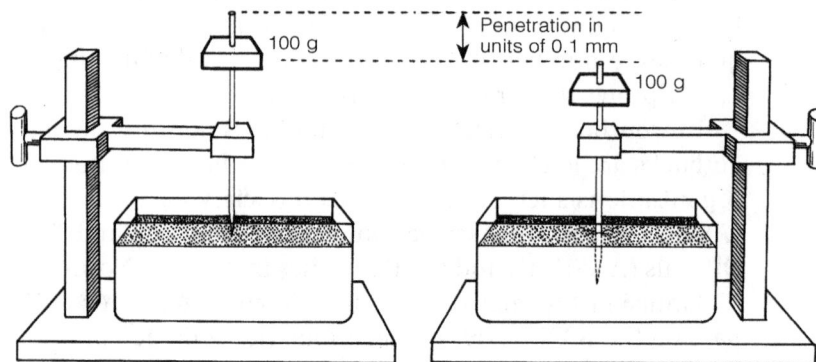

Figure 3.18 The penetration test

Typical values of standard grade bitumens are:
- 40/50 pen bitumen
- 60/70 pen bitumen
- 80/100 pen bitumen
- 150/200 pen bitumen.

Can you arrange the four standard grade bitumens listed above from the softest to the hardest? Think of it as the deeper you are able to stick your finger into a block of softish ice cream the higher the penetration would be.

Viscosity

If a force is applied to a liquid and, in response, it moves quickly, it has a **low viscosity**; if it moves slowly, it has a **high viscosity**. Viscosity is thus a measure of the **flow rate** of a liquid. It is also a measure of the **consistency** (thickness and smoothness) of the liquid.

In deciding on the type of penetration grade bitumen that is required, it is essential to understand what purpose it will be used for. If the bitumen is to be used for road construction, for example, bitumen with a very low viscosity might be too fluid, whereas bitumen with a very high viscosity might be unworkable.

Technological developments have made it possible to test and classify bitumens on the basis of their viscosity, i.e. their resistance to flow or shear. The bitumen's resistance to flow or deformation is governed by its **internal friction**, and can be measured and expressed in units of force required to overcome this friction.

EXAMPLE 1

For an easy experiment to understand viscosity, take a cup of coffee, a cup of honey (or golden syrup) and a cup of bitumen and start to stir each in turn. Notice how easy it is to stir the coffee compared to, say, the honey or the bitumen. The coffee will be termed as having a low viscosity whereas the honey or bitumen will be of a higher viscosity, i.e. its resistance to flow is higher.

Now heat the cups with the coffee, honey and bitumen in them and see if there is any difference when you stir them. You will notice that it now becomes easier to stir both the honey and the bitumen whereas the coffee remains fairly constant. The viscosity of both the honey and the bitumen have changed. It is now lower than before.

Various devices are used to measure viscosity, but in South Africa the viscometer used for specifying penetration grade bitumen is the Brookfield device. Essentially, a sample of penetration grade bitumen is heated to either 60 °C or 135 °C, the spindle inserted into the sample and a torque applied. The liquid's resistance to the torque is then measured, in a unit called a **poise**.

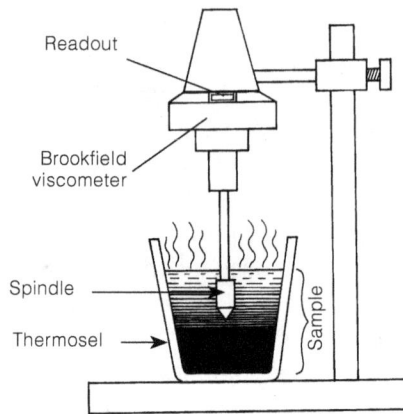

Figure 3.19 Brookfield viscometer

By varying the spindle size, the viscosity can be determined over a large range of bitumen grades – from very viscous or solid to very liquid materials.

Cutback bitumens are also tested in accordance with this test method, as prescribed by ASTM D4402.

The following is an example and an explanation of a typical value obtained in a viscosity test:

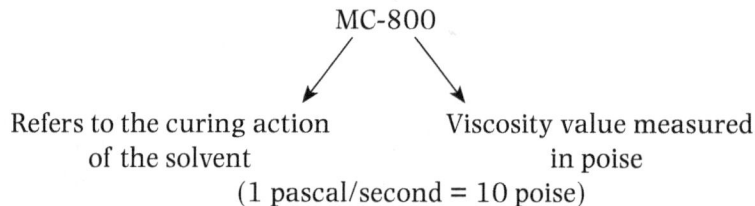

MC-800

Refers to the curing action   Viscosity value measured
of the solvent   in poise
(1 pascal/second = 10 poise)

Typical values obtained are the following:

- MC-30
- MC-70
- MC-800
- MC-3000
- RC-250.

Softening point test

This test is also called the **ring-and-ball test** (R&B test); the name describes the nature of the test apparatus. This test is done by placing a 9.53 mm diameter steel ball on a binder sample placed in a metal ring and immersed in a bath of water. The water is heated until the binder is soft enough to allow the ball to fall to the second metal plate 25 mm below. The water temperature at which this occurs is recorded to the nearest 0.5 °C and is regarded as the softening point of the bitumen sample tested, i.e. the temperature at which the bitumen has a particular consistency.

Take care not to confuse the **melting point** of bitumen with the **softening point**. The softening point is the point at which the bitumen changes from **solid** to **semi-solid**.

This test is usually done in accordance with the method prescribed in ASTM D36.

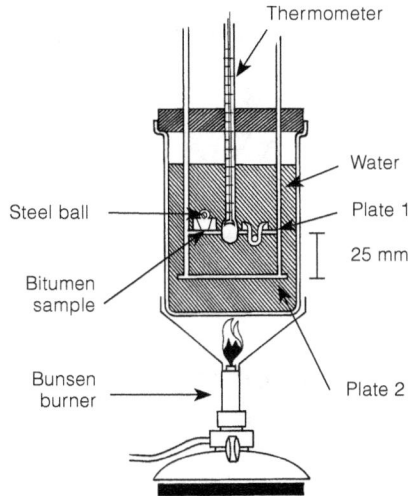

Figure 3.20 The softening point test

Ductility test

What happens with a ductility test?

A ductile material is one which elongates (stretches) when it is subjected to tension and remains elongated prior to breaking. An example is a piece of toffee, which stretches when you pull it and does not return to its original shape or length.

A sample of bitumen with a minimum cross-sectional area of $100\,mm^2$ is pulled at a constant rate until the thread breaks. The distance of this elongation under controlled conditions (usually at a constant temperature of 25 °C) is expressed in centimetres, and this measurement is called the **ductility** of the bitumen.

This test is a measure of the **elasticity** of the bitumen under a slowly applied load. It should be carried out in accordance with the method prescribed in DIN 52013.

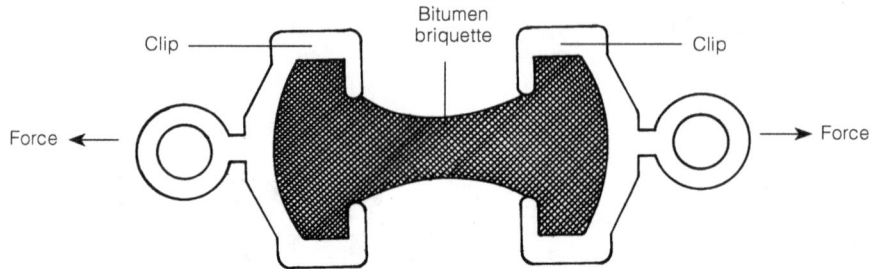

Figure 3.21 The ductility test

3.9.2 Tests on cutback bitumens

Distillation test

The proportion (quantity) and type (quality) of solvent present in a cutback is determined by heating the material, condensing the vapours and noting the volume of the **condensate** collected at various specified temperatures up to 360 °C. Condensation takes place when a vapour changes to a liquid. A condensate is a material produced by condensation. The undistilled portion that remains constitutes the bitumen content of the cutback.

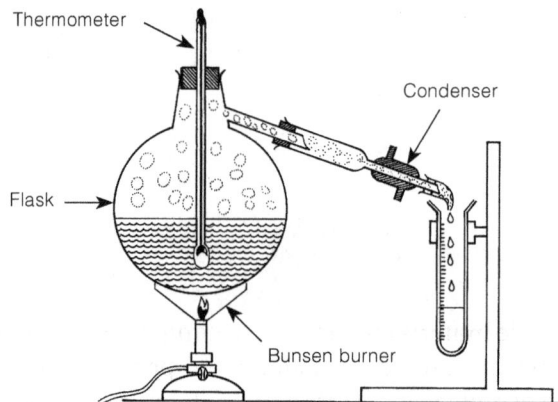

Figure 3.22 The distillation test

Distillation tests can provide useful information about the type of **volatiles** in the binder as well as the rate at which these volatiles will evaporate on site.

The test should be carried out in accordance with the method prescribed in IP27.

Kinematic viscosity test

Kinematic viscosity refers to the force needed to overcome the internal friction in a material, in other words the resistance to flow. Cutback bitumens are classified by their kinematic viscosity at 60 °C, expressed in centistokes (cSt). The type of solvent used, either medium-curing (MC) or rapid-curing (RC), is associated with this. The lower limit of the viscosity range is used in the name of the grade, while the upper limit is double this figure. For example, RC-250 identifies a cutback bitumen of the rapid-curing type which, at 60 °C, has a viscosity in the range 250–500 centistokes.

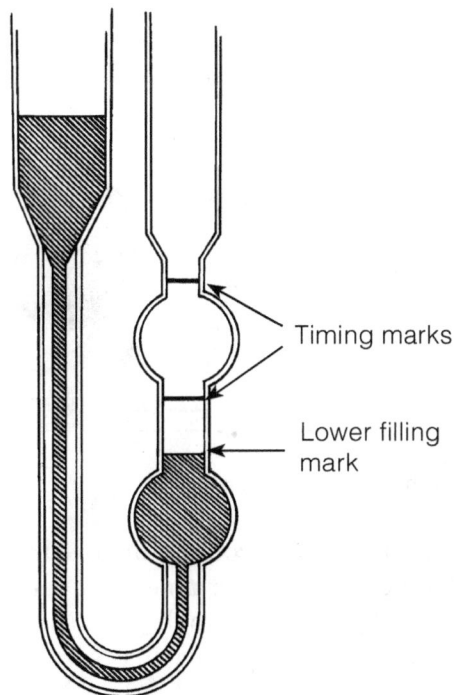

Figure 3.23 The kinematic viscosity test

This test should be carried out in accordance with the test method set out in ASTM D2170

171

3.9.3 Tests on bitumen emulsions

There are several tests that are performed on bitumen emulsions, but you need only know about the binder content test and the particle charge test at this stage.

Binder content test

A distillation procedure is carried out by means of apparatus called the **Dean and Stark apparatus**. An organic liquid that does not mix with water, such as toluene (a petroleum-based liquid used as a solvent) is added to the sample and the flask is heated. The organic liquid distils into the receiving flask, carrying the water with it, and the water then separates into a lower layer. The volume of water is measured and, by difference, the remaining binder content is determined.

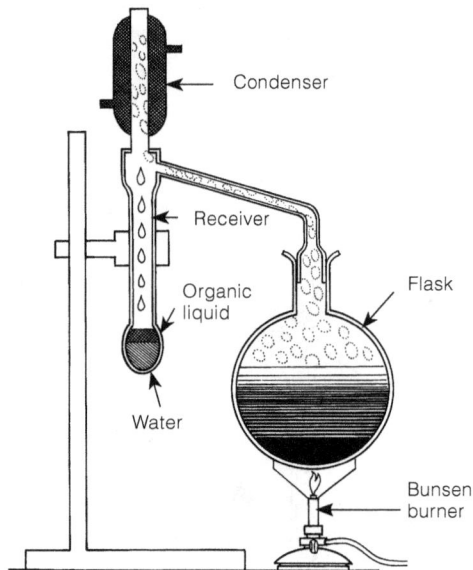

Figure 3.24 The binder content test

Bitumen emulsions may contain up to 40% water by volume, and it is essential to accurately determine the quantity of remaining bitumen that will actually be applied to the road surface.

The test should be performed in accordance with test method ASTM D244.

Particle charge test

The particle charge test is used to distinguish between cationic and anionic emulsions

Immerse two electrodes in a sample of the emulsion and connect them to a low-power, direct-current (DC) source. If, at the end of the specified period, bitumen deposits can be seen on the cathode (i.e. the electrode connected to the negative side of the current source), the emulsion is identified as a cationic bitumen emulsion. If there are no deposits on the cathode, then the emulsion is anionic.

3.10 Applications of bitumen

Bitumen is mainly used in industries and road construction.

In **industries**, blown grade or the harder grades of bitumens are generally used.

Applications include:

- Adhesives
- Felt for roof and floor coverings
- Backing for carpet tiles
- Lining of earth or rockfill dams.

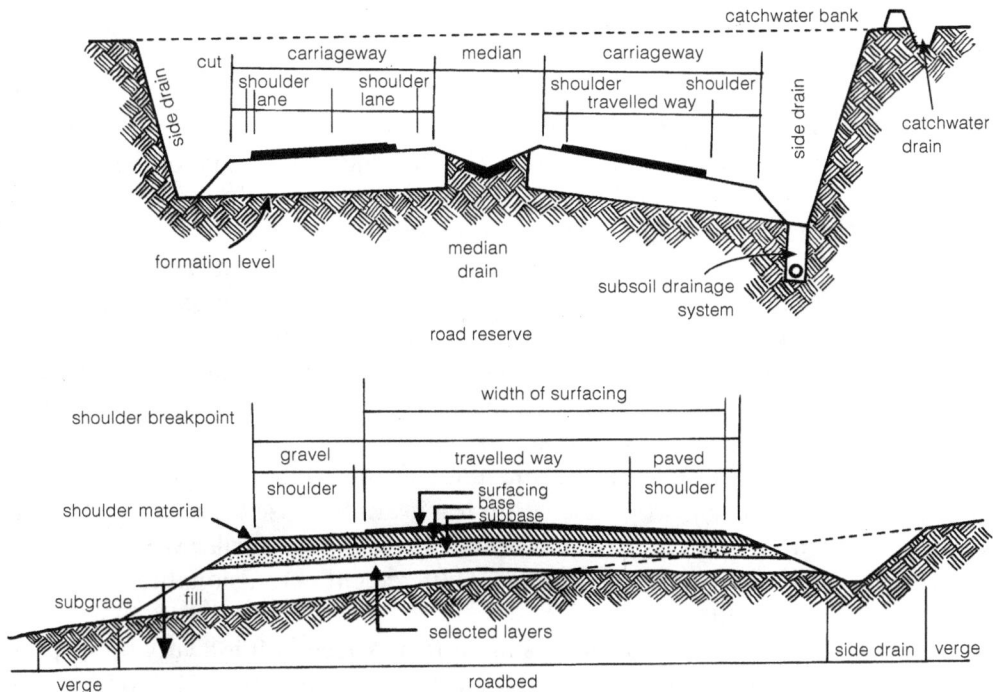

Figure 3.25 Road structure terminology

In **road construction**, bitumen is used in the following ways:

- **Surface dressing:** Where a binder is sprayed on the road surface by a bitumen distributor. An aggregate cover follows immediately, laid by an aggregate reader, and the surface is then rolled as soon as possible. Surface dressing is performed either in single or multiple (two or even three) layers. It is laid on granular bases or existing road surfaces on many types of roads from low-cost surfacing on rural roads to wearing course or surfacing on roads with relatively high traffic. It is important to have a strong base or pavement under the surface dressing. Bitumen, cutback or emulsion is used as binder for surface dressing

- **Slurry seal:** A technique where emulsion, fine aggregates, water and mineral fillers, e.g. cement, are mixed on specially constructed distributors on site. It is used both for preventive or corrective maintenance of road surfaces, such as the sealing of surface cracks. The slurry is usually applied at a thickness of 3–6 mm. The surface is first brushed and then slightly moistened. The mix is laid with the combined mixer-distributor and the curing of the mix starts. Rolling is sometimes done to improve durability

- **Soil stabilisation:** Used to increase the load bearing capacity and firmness of road bases. This operation can be performed using different methods. One method is by blade mixing with a road grader. It is a very simple method but not as efficient as other more advanced methods. The bitumen is applied by a distributor on a **windrow** (a straightish line of material spread across the road) of material. The grader follows immediately and the blade on the machine mixes the material by turning and tumbling actions

- **Tack coat:** A light spray of bitumen done either by hand or machine. It is used to ensure the bond between an old surface and a new asphalt mix layer. A tack coat is usually used when paving. The tack coat must be very thin and cover the entire surface evenly. Too much tack coat may create a slippage plane between the two courses as the bitumen may act as a lubricant. After spraying the tack coat, it must be allowed to 'break' completely before the next layer is placed. During this time, traffic must not be allowed on the sprayed surface

- **Prime coat:** A prime coating is a granular base sprayed over with a binder to prepare for an asphalt surfacing. In earlier years, cutback bitumen was normally used, but recently emulsions have become more common

- **Fog seal:** An application similar to a tack coat. It is a very light spray of slow-setting emulsion with a bitumen content of 30–40 %, either diluted or manufactured. It can be sprayed in one or more layers. Fog seals are used to renew old asphalt surfacings which are, for example, dry, cracked or have surface voids. The emulsion flows easily into the cracks and surface voids and also coats the aggregates in the surface. This method is used to prolong the lifetime of the surfacing and to delay resurfacing

- **Dust binding:** Traffic often creates a lot of dust on unpaved or 'gravel' roads. Spraying bitumen emulsion on the road surface offers a solution to this problem. The emulsion is sprayed using bitumen distributors

- **Minor road repairs:** Sealing of cracks is often done using bitumen emulsions. Cracks appear for several reasons and take many forms from small hairline cracks to major cracks of 20–30 mm wide. Very small cracks are difficult to seal effectively whereas large cracks are filled with emulsion mixed with fine sand. Bitumen and fine aggregate mixes are also used to patch potholes and other damaged areas found in the road surface. Usually smaller quantities are mixed by hand at the work site

- **Recycling:** Increasingly used, this requires fairly specialised machinery that heats up the road surface and then uses tungsten tipped blades to scrape off the upper layer of the road surface (**milling**). The aggregates are then remixed with fresh bitumen and placed back onto the road surface. It is then rolled in (re-compacted)

- **Hotmix asphalt:** Commonly referred to as 'premix', this product is made in a special asphalt plant where both the aggregate (coarse and fine) and the bitumen are heated to temperatures in excess of 160 °C and then mixed together. The material is transported to site using large trucks and then placed with a paver onto the road surface, whereafter it is rolled. A layer of hot-mix asphalt ranges from 20–100 mm in thickness. Usually a penetration grade bitumen, e.g. 40/50 or 80/100, is used

- **Basecourses:** Bitumen is also added and mixed with basecourse material to increase its strength or load carrying ability. Typical bitumen-treated bases are large aggregate mixed bases (LAMBs), granular emulsion mixed bases (GEMs) and emulsion-treated bases (ETBs).

Activity 5
1. When next you walk on a paved road surface, see if you can spot the different cracks as well as whether any crack sealing has taken place.
2. When you see road construction taking place, try to recognise what type of surface application is being applied, e.g. seals, slurries or premix.
3. Although recycling of asphalt is still scarce, if you do see it, stop and ask about the operation.
4. See if you can tell the difference between a tack coat and a prime coat.

5. Have a closer inspection when next you spot road repairs like potholes being done and ask how they are done as well as what material is used.

3.11 Health and safety

Bitumen contains toxic components that have the potential to cause harm to one's health. The health hazards fall into the broad categories of acute and chronic hazards. Acute hazards are associated with single or infrequent events. Chronic hazards refer to regular, repeated events.

3.11.1 Acute hazards

The main acute hazards are associated in the heating of bitumen to temperatures in excess of 100 °C for transportation, handling and/or use. The most common acute hazard is therefore skin burns.

3.11.2 Chronic hazards

These hazards are associated with repeated exposure to bitumen fumes. The major area of concern is the carcinogenic (cancer) potential of bitumen fumes. To date, studies have not provided evidence of a link between bitumen fumes and cancer in humans. The carcinogenic potential of coal tar fumes is, however, well recognised.

Additional hazards are associated with bituminous preparations that contain additives such as aromatics extracts, solvents, distillates and emulsifying chemicals. The required precautions need to be taken when using these products. Repeated and prolonged contact with cutback bitumens can result in skin cancer, and bitumen emulsions can cause skin and eye irritations and possible allergic reactions in some individuals.

3.11.3 Exposure routes

The major routes of exposure are skin contact and inhalation. Additional exposure can occur via ingestion, aspiration and eye contact.

The following emergency treatment is recommended:

- **Skin contact:** No attempt should be made to remove the hot bitumen from the skin. The affected area should be **immersed in cold water** for up to ten minutes. The bitumen will act as a sterile cover over the skin burn and will fall off in time. Where required, only medically approved solvents should be used to remove bitumen from burns. In the case of all burns, medical assistance should be obtained
- **Inhalation:** The patients should be removed to fresh air and their breathing and pulse must be monitored. Give oxygen if necessary and consult the doctor
- **Ingestion:** The patient should not be given anything to drink or eat, and vomiting should be induced or encouraged. If vomiting does occur, keep the air passages clear
- **Aspiration:** If there is evidence of aspiration of cutback bitumen or emulsion into the lungs, seek medical advice immediately
- **Eye contact:** Cool the affected eye under running water for at least 15 minutes. Cold bitumen should be flushed out gently using large volumes of water. Then seek medical advice in all cases.

3.11.4 Areas of risk

The following operational areas are likely to have health and safety risks associated with them:

- **Loading, transportation and discharge operation:** The risks associated with these operations include skin burns, inhalation of fumes and explosions. Where loading of hot bitumen is carried out, care must be taken to ensure that there is no water in the receiving tank, all valves in the receiving tank are closed, fire extinguishers have been removed from the vehicle and are accessible, the vehicle is parked on a level surface, sufficient space is left in the tank for the bitumen to expand when heated and all hatches are securely closed after loading. Where tanks are being cleaned out, extreme caution must be taken in terms of hydrogen sulphide. Prolonged exposure to high concentrations can lead to death
- **Mixing and road paving:** Risks are largely associated with hot products and the resultant fumes. Where the bitumen is diluted with solvents, the risk of skin damage due to skin exposure is greater
- **Surface dressing:** The exposure to fumes occurs in the immediate vicinity of the spraying system. The spray bar and aggregate spreader operators are usually the most exposed. To reduce the health risks, appropriate safety equipment should be worn and operating temperatures kept as low as possible.

3.11.5 Safe handling procedures

The most effective ways to reduce the health risks associated with the use of bitumen products are to:

- **Use personal protective clothing and equipment:** Where hot products are being handled, overalls with close fitting cuffs and leg ends, face and eye shields, gloves and boots must be worn. Due to the high degree of manual handling that takes place, it is vital that employees are given correct information. This is also a requirement of the Health and Safety Act
- **Control temperatures to reduce fume emissions:** Bituminous products should be stored and handled at as low a temperature as possible. Where solvents are being used, refer to the proper handling procedures
- **Improve personal hygiene:** Where the risk of skin contact exists, it is essential to adopt a high standard of personal hygiene. This includes provision of washing facilities without hot water and soap, using non-solvent based skin cleaners, skin conditioners, providing clean overalls, providing separate storage areas for personal and working clothes, ensuring that clean gloves are used and old ones disposed of, etc. Regular monitoring of the work environment, hygiene procedures and state of safety clothing should be carried out.

Self-evaluation 3.2

1. Complete the sentences:
   a. Bitumen tests can be categorised according to _____ or _____.
   b. A low penetration value will indicate a _____ bitumen.
   c. A _____ _____ apparatus is used to determine the viscosity of bitumen.
   d. A ductile material _____ when in tension.
   e. Hot mix asphalt (HMA) is a blend of _____ and hot _____.

2. State whether the following statements are **true** or **false**:
   a. Consistency relates to the penetration of a bitumen.
   b. During the standard penetration test, a 5 gram needle is allowed to penetrate the sample for 100 seconds.
   c. During the ring-and-ball test, the water temperature is recorded to the nearest 0.5 °C.
   d. The unit of measurement that is used in the kinematic viscosity test is Pa.s.
   e. The Dean and Stark apparatus is used for the viscosity tests on bitumen emulsions.

    f.   The particle charge test distinguishes between cationic and anionic bitumen emulsions.
3. Answer the following:
    a.   Why is it necessary to conduct tests on bitumen?
    b.   What are we trying to establish by conducting the following tests?
- Viscosity test
- Distillation test
- Particle charge test
- Penetration test.
    c.   How would you interpret the following bitumen test results?
- 150/200 penetration bitumen
- MC-3000 cutback bitumen.
    d.   List the various applications of bitumen in the road industry.

3.12 Summary

The purpose of this unit was to:
- Introduce bitumen, which is widely used in the civil engineering industry
- Discuss the production of the different types of bitumen, their composition and characteristics
- Explain sampling methods, safety precautions and some of the tests carried out on bitumen to ensure a quality product
- Show that bituminous products are mainly made up of hydrocarbons and have certain common properties such as viscosity, adhesion and durability
- Give an overview of bitumen's wide range of applications.

Answers

Activity 1
1. Carbonisation does mean to burn, but it involves applying very intense heat that can only be obtained during the smelting process in industrial plants, which then breaks coal down to carbon from which tar is manufactured.
2. Fuel is a wide term used to mean anything that is burnt to provide heat or power. Different types of fuel are: liquid fuels, e.g. petrol used in motor vehicles and paraffin in stoves; solid fuels, e.g. coal used in steel plants and electrical substations; fuel in gaseous form, e.g. space rocket fuel; synthetic fuels are fuels that are not

found in nature, but are manufactured by a chemical process; petroleum is a term Americans use, but we call it crude oil because it is oil as it is found in nature.

Activity 2

1. Refresh your memory by reading section 3.2 and section 3.4.
2. See table 1. 1.
3. Read section 3.5 to familiarise yourself again. Hint: you should be looking at aliphatics, cyclics and aromatics.
4. Once again section 3.5 will do the trick. Hint: look at the paragraphs on maltenes and asphaltenes.
5. Read up on the five points under section 3.6.
6. An explanation of a binder is given in the unit introduction. A binder is any kind of bitumen-based material that binds particles together. Bitumen is thus one kind of binder.
7. – Hydrocarbons are molecules made from a combination of hydrogen and carbon atoms.
    – The viscosity of a material is the material's resistance to flow.
    – Viscoelasticity is a behaviour that depends on temperature. At high temperatures, bitumen can be very liquid and at low temperatures, it can be very stiff. It is, therefore, a visco-elastic material.
    – Tar is a by-product of the burning of coal and is also made up of complex hydrocarbons.
8. – Aliphatics have an oily or waxy composition.
    – Cyclics are three-dimensional with various atoms attached.
    – Aromatics are flat, stable carbon rings stacked together.

Activity 4

1. See section 3.8 and table 3.3.
2. Read the items under section 3.8.5 to refresh your memory.
3. Read section 3.8.3 where it deals with the procedures. You need to separate them out for bulk or drum sampling.
4. Your answer should indicate that sampling is essential to ensure that you will be using the correct type of bitumen.
5. Such documentation will help to ensure that people comply with international standards and control measures for manufacturing the materials, so that everyone knows what performance to expect from the product.
6. Either through a thief sampler or sampling valve.
7. If they are contaminated, then you will not have a true reflection of the material on site.
8. So that we can identify them.
9. See section 3.8.4.

10. Four drums, because the cube root of 64 is four.
11. The precautions that should be taken are set out in section 3.8.5. They are necessary to protect the people working with the samples, as well as the environment.

Self-evaluation 3.1

1. a. Hard; soft; intermediate
    b. Consistency; penetration; viscosity; softening point
    c. Cutter; flux
    d. Rubbery
    e. Invert
    f. Negatively
    g. Evaporation
2. a. False, penetration grade bitumens are classified according to penetration
    b. True
    c. False, there are three, not two
    d. True, RC stands for rapid-curing; MC for medium-curing; and SC for slow-curing
    e. True
    f. False, the first phase is the water phase
    g. True
3. a. Section 3.7 and its subheadings deal with each of these types of bitumen separately.
    b. Part of the answer to this question is addressed in section 3.7.4 but the diagram is not. You will therefore have to go to the library and take out a book on emulsions to get the diagram or you can look in Sabita Manual 2.
    c. Once again section 3.7.4 deals extensively with the answer to this question. Things to look at are temperature, viscosity and safety of emulsion compared to other bitumens.
    d. Read section 3.7.5. Hint: there are basically two types, i.e. homogeneous and non-homogeneous and under each are subdivisions or products, e.g. SBS, SBR, rubber crumbs, etc.

Activity 5

Compare your observations to what you have studied.

Self-evaluation 3.2

1. a. consistency; composition
    b. hard
    c. Brookfield viscometer
    d. stretches
    e. aggregate; bitumen

2. a. True
    b. False, a 100 gram needle is applied for 5 seconds
    c. True
    d. False, it is measured in centistokes
    e. False, it measures the amount of binder in the emulsion
    f. True

3. a. To ensure that the product is within certain specifications, or as a means of quality control.
    b. – Penetration test – to determine the hardness/softness of the bitumen
        – Viscosity test – to measure the resistance to flow of the liquid or to measure the internal friction of the liquid.
        – Distillation test – to determine the amount and type of solvent present in the cutback bitumen.
        – Particle charge test – to identify whether the emulsion is positively or negatively charged (whether it is cationic or anionic).
    c. – 150/200 pen bit means that during the test the needle has penetrated the bitumen sample with an upper limit of 150 mm and a lower limit of 200 mm, indicating that it is a fairly soft bitumen.
        – MC 3000 is a medium-curing cutback bitumen with a viscosity measured as 3 000 centistokes.
    d. See section 3.10 for a list of applications.

Advanced exercises

1. Bitumen has a tendency to be temperature susceptible. What do you understand by this term and what is the effect of this phenomenon when used in roads?
2. What do you understand by the term 'ageing of bitumen'?
3. Evaluate the use of either concrete or bitumen when used in roads, based on the variables influencing their use as well as the following criteria:

Quality of performance	Maintenance
Functionality (rideability, noise, etc.)	Workability and ease of construction
Design	Energy
Economy + cost per m^2	Safety

4. When a fluid flows easily, is it deemed to have a high or low viscosity?
5. Why do we need to know or measure the viscosity of bitumen?
6. Why is it necessary to modify binders?

7. Is there a relationship between temperature and viscosity (direct or indirect)? If so, what is it and why?
8. Is there a relationship between viscosity and movement of liquid? If so, what is it?
9. Is there a relationship between temperature, viscosity and the movement of liquid? Sketch this relationship.

4

Structural materials: brick, timber, steel and aluminium

Learning outcomes

After studying this unit, you should be able to:

- Describe the manufacturing, characteristics and properties of a number of different materials like bricks, steel, aluminium and timber
- Describe defects in materials
- Calculate the quality of bricks in a wall
- Distinguish between different types of timber products, metals, steels, aluminiums and their uses
- Apply the relevant standards and codes.

4.1 Introduction

In this chapter you will be guided through the everyday materials found around you and those that you will encounter in civil engineering. Sometimes these materials are taken for granted just because they are there and available but yet they form essential parts in any construction project. Figure 4.1 shows how some of these materials are used in construction. Replace this image with a high-rise or multi-storey building and you will find that the same principles and materials apply, many are just used in another way.

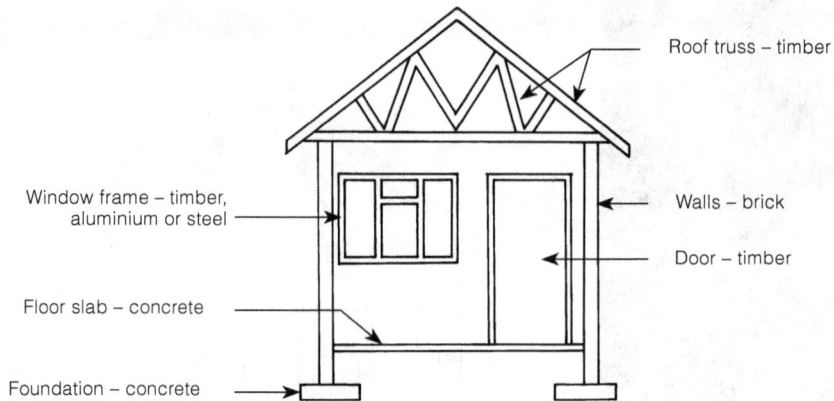

Figure 4.1 A house construction

4.2 Bricks

Bricks are probably the oldest industrialised building material. The earliest bricks were made from clay, taken from sources close to the surface of the ground or from river banks and moulded into shape by hand and dried in the sun. People from the Bronze Age realised that firing (setting fire to or burning) the brick-shaped clay resulted in greater permanence or hardness. By the time the Romans arrived on the scene, the art of burning the moulded brick for durability was well known and practised.

Clay bricks were traditionally made locally and not transported very far, so that they had widely differing characteristics depending on the material available and the way the bricks were treated by the maker. In the 19th century, a process was developed for making bricks from sand and lime by subjecting them to high-pressure steam to bind the materials together by the formation of calcium silicate. These days bricks are also made from concrete. Concrete bricks are made using the normal materials associated with mortar, i.e. sand, fine aggregate and cement to which colouring agents may be added.

Figure 4.2 An early brick-making site

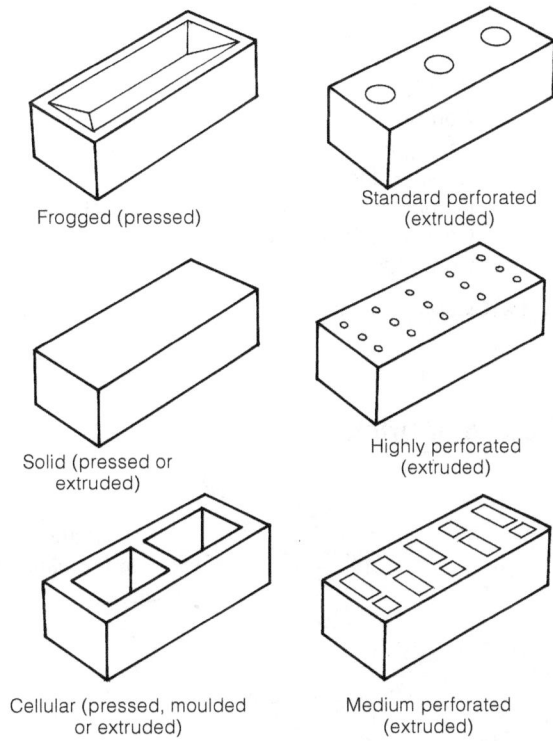

Frogged (pressed)

Standard perforated
(extruded)

Solid (pressed or
extruded)

Highly perforated
(extruded)

Cellular (pressed, moulded
or extruded)

Medium perforated
(extruded)

Figure 4.3 Different brick types

4.2.1 Types of brick

> *Bricks are classified by quality and manufacturing process.*

Bricks are classified both by their **quality**, which affects their use, and the **manufacturing process**, which affects their appearance (see fig 4.4).

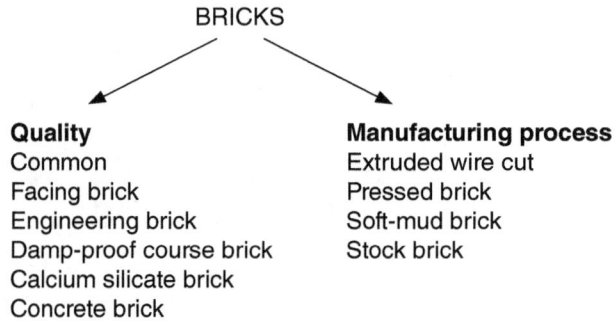

BRICKS

Quality
Common
Facing brick
Engineering brick
Damp-proof course brick
Calcium silicate brick
Concrete brick

Manufacturing process
Extruded wire cut
Pressed brick
Soft-mud brick
Stock brick

Figure 4.4 Classification of bricks

4.2.2 Manufacture of bricks

> *What are the four main stages in the manufacture of bricks?*

There are four main stages in the manufacture of clay and shale bricks:

- Obtaining and preparing the material
- Shaping
- Drying
- Firing.

When a plastic clay with high moisture content is used in the process, there is usually an intermediate stage of drying between the shaping and the firing processes. For facing bricks, the finished appearance may depend upon additional work at the shaping stage, usually either by application of coloured sand to the surface of the wet clay or by mechanical texturing of the surface. In some processes other materials are added to clay, either to improve appearance or to incorporate combustible materials to assist firing.

Clay preparation

After the clay has been extracted from the pit, it is prepared and mixed in a variety of ways, depending upon the type of raw material and particular requirements of the subsequent shaping process. Water content is controlled so that the material going on to the shaping process may vary from very wet to relatively dry clay dust. In most cases the clay contains some traces of iron oxide that will give bricks a reddish colour.

Figure 4.5 Excavating clay material for brick-making

Shaping the bricks

There are seven shaping systems. One is an extrusion process, the others all involve materials being forced into moulds by pressure.

In the extrusion method, holes of a variety of shapes may be formed through the bricks; these are known as perforations. In the moulding processes, indentations such as frogs (see fig 4.6) and hollows may be formed.

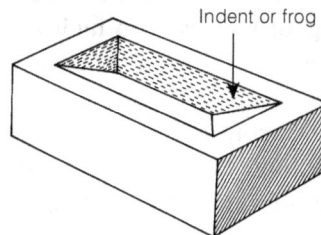

Figure 4.6 A frog in a brick

The shaping method used depends upon the type of raw material and in particular upon water content and type of brick required.

Hand-made bricks are made by rolling a lump of high quality, very plastic clay in moulding sand and throwing it into a mould by hand. Although more expensive than machine-made bricks, this method gives a particularly attractive surface finish that is difficult to reproduce mechanically.

Soft-mud bricks are manufactured by machine or by hand with natural clay or with a mixture of clay and lime. The bricks are usually frogged and less accurate in shape than other forms of brick. Sand is usually used in the moulds to enable the bricks to be easily removed and this causes an uneven patterning or creasing on the face.

Figure 4.7 Producing handmade bricks

Semi-dry bricks are made from a fine clay dust that is delivered to machines, called pugmills, where it is pressed to shape in moulds. The pressing process is usually repeated, up to four times, to get consistency of the shape and mix. Because the material is relatively dry the shaped bricks go direct to the kilns for firing, without intermediate drying.

Bricks made by this process are fairly regular in shape and size but the commons are unattractive. When intended for facing work they may be either sand faced or machine textured after the pressing process. These surface finishes are often noticeably different in colour from the body of the brick and care is needed in delivery and site handling to avoid unsightly damage.

Pressed bricks may have one or two frogs or may have a variety of other shaped holes pressed in to produce the class of brick described as cellular.

Stiff-plastic bricks are made by a similar process to semi-dry bricks but because the clays and shales used have relatively low plasticity, some water is added to the clay dust before the material is delivered to an extrusion pug which forces roughly brick-sized clots into moulds. A press die gives the final shape and compaction to each brick. The water content is low enough for the bricks to go direct from pressing to kiln.

Wire-cut bricks use clay that is usually fairly soft and of fine texture. It is extruded as a continuous ribbon and cut into brick units by tightly stretched wires spaced at the height or depth for the required brick. Allowance is made during the extrusion and cutting for the shrinkage that will occur during firing. Wire-cut bricks do not have frogs and on many the wire cutting marks can be clearly seen. The wire-cut process provides flexibility for producing varying sizes and shapes of brick and is an economical method for mass production.

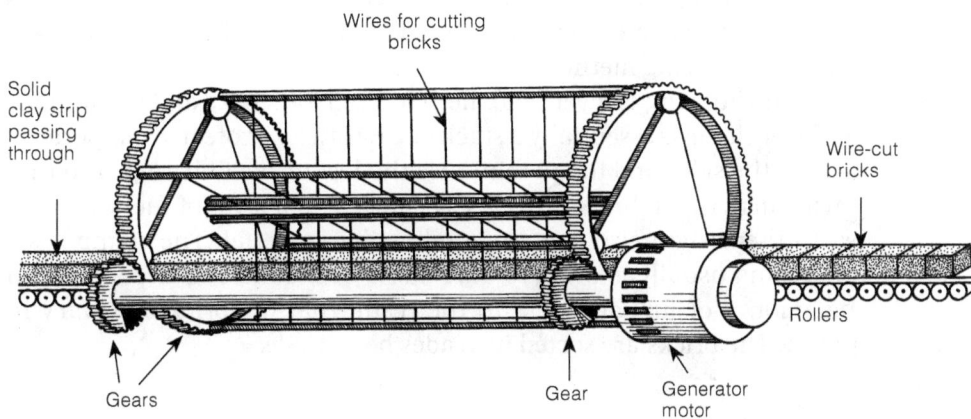

Figure 4.8 The wire-cutting process

Facing and engineering bricks are sometimes pressed after being wire cut to provide smoother faces and sharper arrisses (see section 4.2.10). Holes formed during the extrusion process vary considerably in number and size. Bricks with total volume of holes less than 25 % of volume of brick are called **perforated bricks.**

The advantages of perforated bricks include a reduction in process times, reduction in weight and some increase in thermal value. The perforations do not appear to affect rain penetration through walls. As the holes go right through the bricks when laid vertically, they are vulnerable to saturation from rain during construction. Saturation may increase drying-out time and may inhibit early decoration, such as painting, due to wet walls.

Drying

Drying is necessary in all cases where the brick, after forming, is soft and unable to withstand the weight of other bricks when stacked for firing. This is the case with all hand-made and soft-mud types and also with wire-cut bricks where the moisture content is relatively high. Drying is carried out in a series of chambers or tunnels in which the bricks are arranged so that a flow of heated air can pass over them. The temperature and humidity of the air is regulated to control shrinkage which takes place during drying.

Firing

Firing produces a number of complicated chemical and physical changes in clay and the degree of control obtained is important. The ultimate firing temperature and type of atmosphere affect the colour of the bricks. Temperatures reach to 1200 °C inside the kiln. The fuel used can vary from grass and litter to more commercially available options like heavy furnace oil, gas, powdered coal and anthracite. There are four types of firing methods:

Clamp burning is a very old method used for stocks and hand-made bricks. A clamp is essentially a stack of raw bricks, protected temporarily around the sides and top to minimise the loss of heat, arranged so that it will burn. Green bricks are closely stacked on a layer of fuel while the bricks themselves also contain combustible material. The clamp is set on fire and usually allowed to burn itself out. Control is dependant on the amount of fuel added, with the result that the brick may vary in quality. The bricks are sorted in grades before sale.

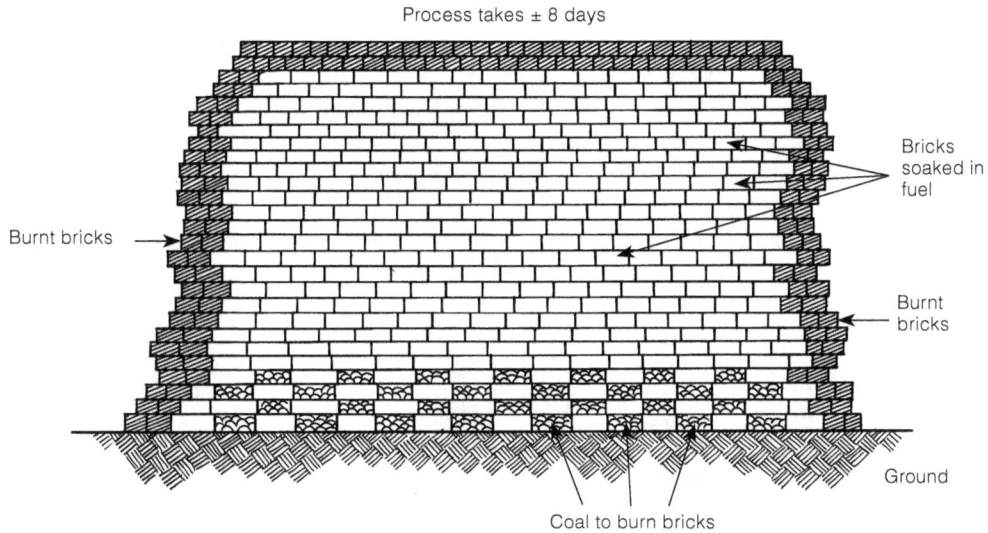

Figure 4.9 Clamp burning

Intermittent kilns are also a very old method. These and clamps were the two methods used before the continuous kiln was introduced in 1858. They are now used for the manufacture of special types of bricks. The kiln is a permanent structure but the process is not continuous. After the bricks are loaded in the kiln, they are burnt, cooled and unloaded. The kiln is then prepared for the next batch. There is no wastage of fuel, but the bricks on the top of the stack may be underburnt while those at the bottom may be overburnt.

Continuous kilns consist of a number of chambers connected so that the fire can be led from one chamber to the next so that the stationary bricks are heated, fired and cooled. The system combines economy and a good degree of control.

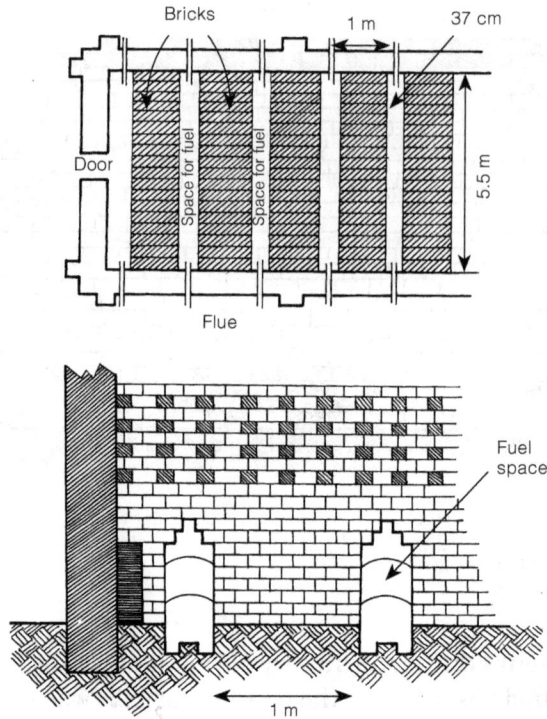

Figure 4.10 An intermittent kiln

However, the bricks need to be moved quickly before the fire catches up to the chamber so as not to disrupt the firing sequence. Because the bricks are stationary and the fire is moved from chamber to chamber, it is important that a properly managed burning process is adopted when using this type of kiln, otherwise the bricks could be burnt more than once.

Tunnel kilns are a popular recent innovation. In tunnel kilns the fire remains stationary while the bricks, carried on kiln cars, pass along a tunnel through preheating, firing and cooling zones. The ability to vary temperatures and track speeds provides optimum conditions for quality control.

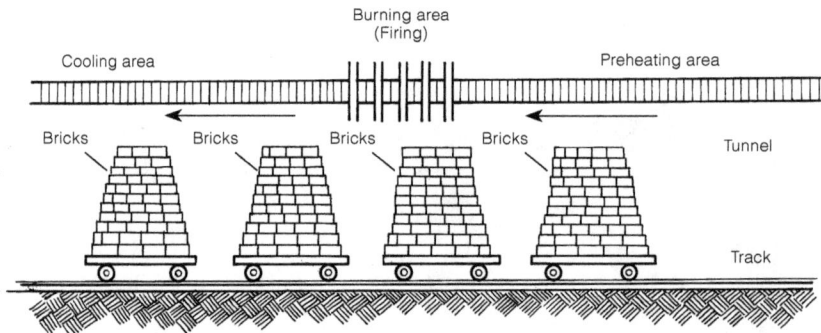

Figure 4.11 A tunnel kiln

4.2.3 Clay bricks

How many types of clay bricks are made in South Africa?

There are **three basic types of clay bricks** produced in South Africa.

Non-face plaster bricks (NFP) are used on both internal and external walls. They require some form of covering, usually plastering or tiling. They are also called commons, stock bricks, common bricks or run-of-kiln (ROKs) bricks.

Face bricks are produced to withstand most environmental conditions as well as provide an aesthetic appearance. They are sub-divided into three categories:

- **Face brick aesthetic** are carefully selected for their special beautifying character like shape, size and colour, e.g. rock-faced bricks and rustics
- **Face brick standard** are durable and uniform in size and do have a beautiful appearance, but not quite in the same category as face brick aesthetic (FBA), e.g. smooth satin-textured bricks
- **Face brick extra** possess the highest degree of size, shape and colour uniformity. They are used where detailed architectural brickwork is required.

Non-face extra bricks do not require any form of covering, but are used under damp conditions, especially in foundations. They are commonly referred to as footing bricks, foundation bricks or hard burnt commons.

4.2.4 Calcium silicate bricks

The raw materials used in calcium silicate bricks are lime, silica sand and water. The raw materials are mixed in carefully controlled proportions with colouring materials added as required. The material is shaped by pressing and the bricks are then loaded onto bogies and passed into autoclaves where they are steamed under pressure. By varying autoclaving time and steam pressure, the performance characteristics of the bricks can be adjusted. Colour and quality are determined by the mix and the autoclaving process.

4.2.5 Concrete bricks

Concrete bricks are made from a carefully controlled mixture of cement, sand and aggregate together with additives such as colouring agents. The mixture is pressed and/or vibrated into brick-sized moulds in much the same way as calcium silicate bricks. The moulded bricks are then cured either in steam chambers or in the air. The product is regular in size and colour.

4.2.6 Concrete blocks

Concrete blocks are similar to concrete bricks except the blocks are larger, usually 390 mm × 190 mm × 90 mm (length × width × height). Concrete blocks are used for both internal and external walls. It is often recommended that they are plastered to provide extra protection against moisture. Blocks are generally used in low-cost housing because they are less expensive than clay bricks and speed up construction.

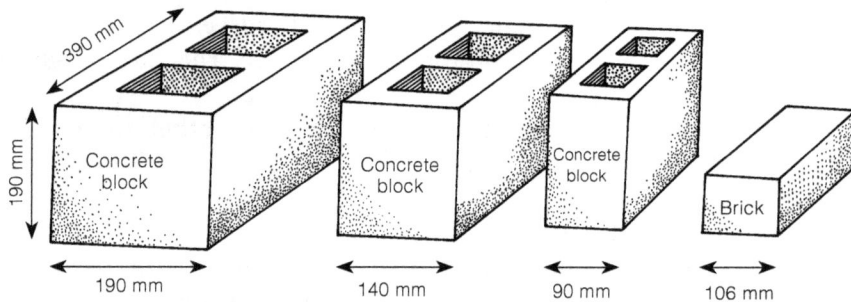

Figure 4.12 Dimensions of concrete blocks and bricks

4.2.7 Properties of bricks

Can you list the properties of bricks?

The properties against which bricks are measured are:
■ Size
■ Water absorption

- Compressive strength
- Soluble salt content
- Frost resistance
- Efflorescence
- Thermal movement
- Moisture movement.

Size varies depending on manufacturing process, but the usual standard for clay bricks is 222 mm × 106 mm × 73 mm. When describing the dimensions of a brick, the standard procedure is length × width × height (all in millimeters).

Water absorption is the % increase in the weight of a dry brick when it has been saturated. It is one of the parameters for the definition of engineering bricks and damp-proof course bricks. It affects rain penetration through the outer skin of a cavity wall and is used to define the flexural strength in lateral load design. The water absorption test requires a brick to be soaked in a bath of water for 24 hours to determine its ability to absorb water. Depending on the type of brick, this will range from 5% to 15%. The more porous the brick, the more water it will absorb.

The **compressive strength** of a brick is the mean of 10 crushing tests in which the failing load is divided by the gross area of the brick. The compressive strength of manufactured clay bricks can vary between 3.5 MPa and 49 MPa. A guaranteed compressive strength of 17 MPa is, however, specified for facings. Engineering bricks can withstand much higher loads than ordinary common bricks.

$$\text{Compressive strength} = \frac{\text{load at failure (F)}}{\text{area of the brick (A)}}$$

When undertaking the test, the brick must first be immersed in clean water for 24 hours. The brick is then placed between the plates of a compression testing machine and a load gradually applied until the brick fails. The maximum load is determined and averaged out for the 10 tests done.

Soluble salt content

Most clays used in brick-making contain soluble salts that may be retained in the fired bricks. If brickwork becomes saturated for long periods, soluble sulphates may be released. This sulphate attack may cause mortars that have been incorrectly specified or batched and have a low cement content to deteriorate.

Frost resistance
The frost resistance of calcium silicate bricks is generally higher than ordinary bricks and they should not be used where they may be subject to salt spray.

Efflorescence
Efflorescence is a crystalline (whitish) deposit left on the surface of clay brickwork after the evaporation of water carrying dissolved soluble salts. Efflorescence is harmless and usually temporary. It can be minimised by protecting the bricks from rain at the early stage of construction.

To test for efflorescence, a brick is immersed 25 mm deep in water and left in a well-ventilated room at a temperature of 20–30 °C until all the water evaporates. This process is then repeated. After the second evaporation has taken place, the brick is examined for white patches. According to SANS 227: Standard Specification for Burnt Clay Masonry Units, efflorescence can be classified into five degrees:
- **Nil:** No perceptible deposit of salt
- **Slight:** A very thin deposit of salts, just perceptible, or a small amount of salts occurring only on the edges of the masonry
- **Moderate:** A heavier deposit than 'slight', but has not caused powdering or flaking of the surface
- **Heavy:** A thick deposit of salts covering a large area of the unit, but that has not caused powdering or flaking of the surface
- **Serious:** A deposit of salts that has caused powdering or flaking of the surface.

Thermal movement
Bricks expand on being warmed and shrink when cooled. It is the expansion and contraction of the brickwork as a whole and not a single unit that is of interest to a civil engineer.

Moisture movement
Clay bricks expand on cooling from the kiln as some of the water molecules re-attach themselves after being driven off by the heat of the kiln. This expansion is non-reversible unless the bricks are re-fired. The magnitude of this movement varies according to the type of brick.

Activity 1
You have now encountered some of the properties of clay bricks, but have you ever held one in your hand and tried to estimate how much it weighs? Get hold of some of the different types of bricks (clay and concrete) and write down what you think the mass of each is. Then measure their mass on a laboratory scale. Compare the results with your estimates to see how close you were.

4.2.8 Defects in bricks

What are the principal defects in bricks?

The following are the principal defects to which bricks are subjected:

- **Black core** or **hearting** is common in bricks made from red clay which have been heated too rapidly in the kiln, causing the surface to vitrify and the interior to remain black
- **Bloating** or **swelling** is caused by an excess of carbon matter present in the clay and due to bad burning
- **Burring** or **clinkering** happens when damp burnt bricks, usually adjacent to the flues, have been fused together by excessive heat. These bricks are called clinkers. These bricks, if not broken up for coarse aggregate, are used as architectural features. This clinkering must not be confused with the clinker you learnt about during the manufacturing phase of cement
- **Chuffs** or **shuffs** are badly cracked and mis-shaped bricks, produced by rain falling on them when they are hot
- **Crazing** is common in glazed bricks and is usually shown up by fine cracks, caused by the glaze and clay not expanding and shrinking to the same extent
- **Efflorescence** occurs in bricks made from clay containing a large proportion of soluble salts, which are liable to become discoloured by the formation of whitish deposits on their surface. It is common in old and new bricks. The salts are dissolved when water is absorbed by the brick. As the brick dries, the salt solution is brought to the surface, evaporation takes place and the salt remains on the face
- **Grizzles** are common bricks which are underburnt and therefore weak (the process is known as grizzling). They are usually recognisable by a light colour and a dull sound when struck. They are suitable for internal partitions where little strength is required
- **Iron spots** are dark spots on the surface caused by the presence of iron sulphide in the shale, which will render the bricks unsuitable for facings.

4.2.9 Size of bricks

Uniformity in the size of bricks is essential if the maintenance of the correct bond is to be kept (see sectiom 4.2.11) during the construction of the wall. Time is wasted if a consignment contains bricks of varying sizes, as the bricklayer is then required to make a selection as the work proceeds. Bricks in general are approximately 222 mm long × 106 mm wide × 73 mm high.

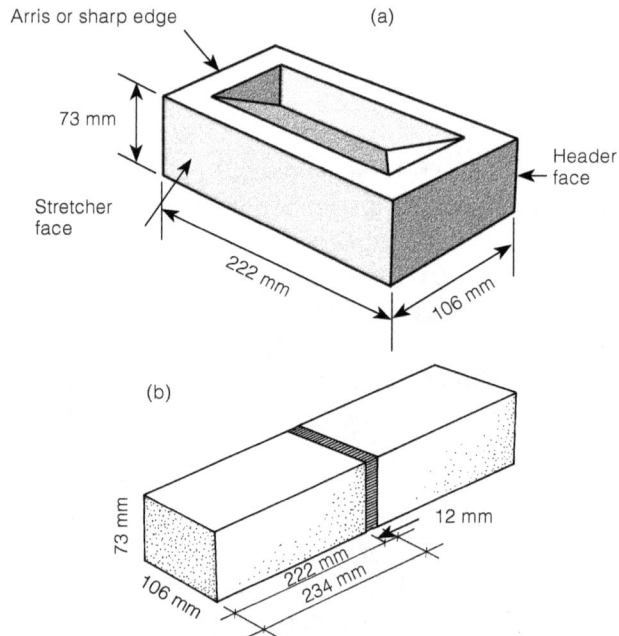

Figure 4.13 (a) Dimensions of a brick (b) Dimensions of a brick plus mortar joint

4.2.10 Terms used in brick laying

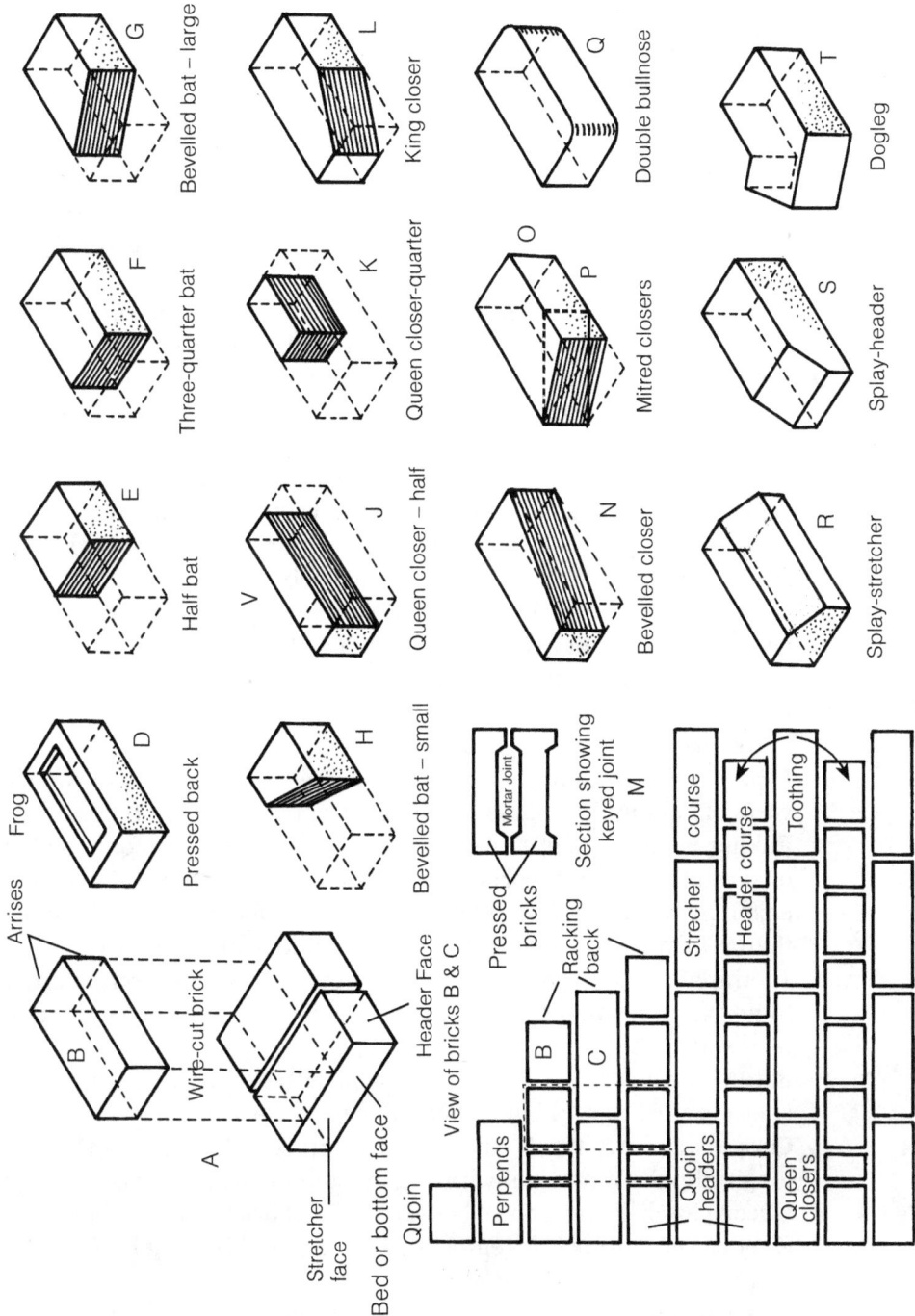

Figure 4.14 Terms used in brick laying

- **Arris:** An edge of a brick
- **Bed:** The lower 222 mm × 106 mm surface of a brick when placed in position
- **Header:** The end or 106 mm × 73 mm surface of a brick
- **Stretcher:** The 222 mm × 73 mm side of a brick
- **Face:** A surface of a brick, such as the header face and stretcher face. The term is also applied to an exposed surface of a wall
- **Frog or kick:** A shallow sinking formed on either one side or both of the 222 mm × 106 mm face of a brick. A wire-cut brick has no frogs, a pressed brick has two frogs as a rule. A frog affords a good key for the mortar
- **Bed joints:** Mortar joints parallel to the beds of the bricks, and therefore horizontal in general walling, thickness from 3–12 mm
- **Course:** A complete layer of bricks plus its mortar bedding joint. A header course consisting of headers and a stretcher course comprising stretchers
- **Brick gauge:** The height of a brick course
- **Continuous vertical or straight joints:** Vertical joints which come immediately over each other in two or more consecutive courses. Although these are sometimes unavoidable (Flemish bond), they should not appear on the face of the brickwork (English bond)
- **Quoin:** A corner, or external angle of a wall
- **Stopped or closed end:** A square termination of a wall, as distinct from a wall which is returned
- **Perpends:** Imaginary vertical lines which include vertical joints
- **Lap:** The horizontal distance which one brick projects beyond a vertical joint in the course immediately above or below it
- **Raking back:** The stepped arrangement formed during the construction of a wall when one portion is built to a greater height than the adjoining. No part of a wall during its construction should rise more than 900 mm above another if unequal settlement is to be avoided
- **Toothing:** Each alternate course at the end of a wall projects in order to provide adequate bond if the wall is continued horizontally at a later stage
- **Bat:** A portion of an ordinary brick wall with the cut made across the width of the brick (½ bat, ¾ bat and bevelled bat)
- **Closer:** A portion of an ordinary brick with the cut made longitudinally, and usually having one uncut stretcher face
- **Queen closer:** Usually placed next to the first brick in a header course
- **King closer:** Formed by removing a corner and leaving half header and half stretcher face bevelled
- **Bevelled closer:** Has one stretcher face bevelled
- **Mitred closer:** Used in exceptional cases as and when the ends are required to be mitred

- **Bull nose:** Used for copings, or in such positions where rounded corners are preferred to sharp arrises
- **Dog leg or angle:** Bricks used to ensure a satisfactory bond at quoins which depart from a right angle. Preferable to mitred closes.

4.2.11 Types of bond

There are many varieties of bond, the most common being:

- **Header,** where bricks are placed with the headers facing the outside

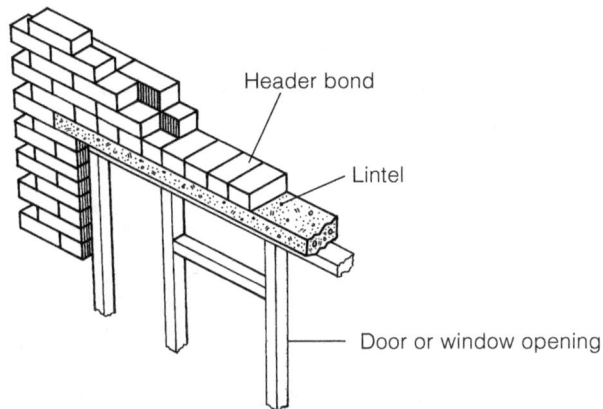

Figure 4.15 A header bond

- **Stretcher,** where bricks are placed with their longest side facing the outside of the wall

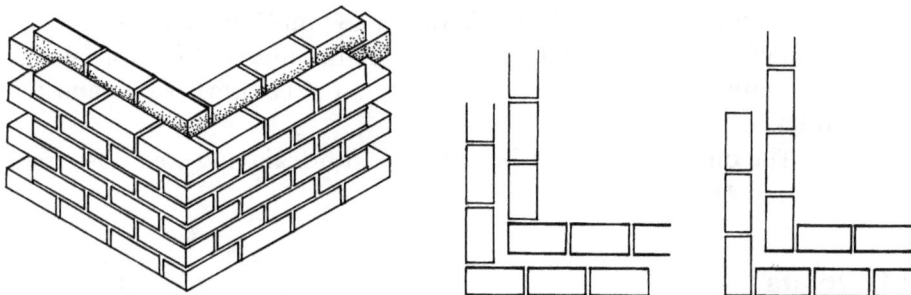

Figure 4.16 A stretcher bond

- **English bond,** which consists of alternate courses of headers and stretchers

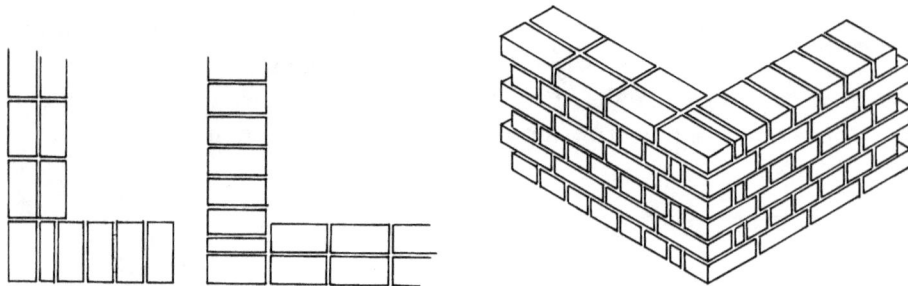

Figure 4.17 An English bond

■ **Flemish bond**, which comprises headers and stretchers in each course. It has a better appearance than an English bond, but is not quite as strong.

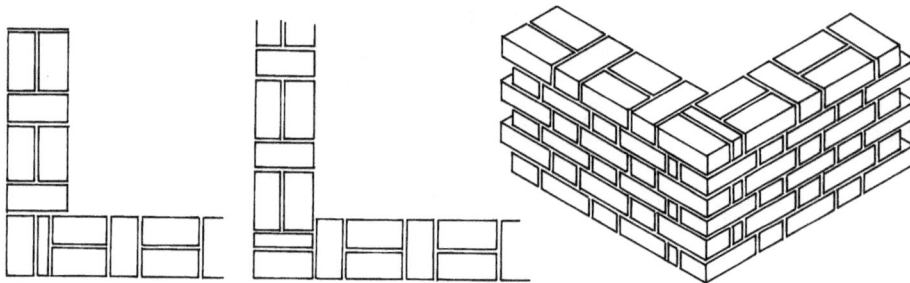

Figure 4.18 A Flemish bond

The thickness of walls is either expressed in millimeters (mm) or in terms of the width of the brick, thus 118 mm (106 mm + 12 mm joint) or half brick, 234 mm (222 mm + 12 mm mortar joint) or one brick, 356 mm (222 mm + 12 mm + 110 mm + 12 mm) or one-and-a-half brick, etc.

The different types of bonds used in bricklaying were discussed in section 4.2.11.

4.2.12 Joints

There are various types of joints used in the construction sector. Generally these joints can vary between 3–12 mm depending on the type of brick used, the construction method or the aesthetic appearance.

4.2.13 Calculating quantities in a wall

Bricks

To calculate the quantity of bricks in a wall, assume a standard brick size to be 222 mm × 106 mm × 73 mm and also allow for a mortar joint of 12 mm. The size of the stretcher face would then be:

222 mm + 12 mm = 234 mm, and
73 mm + 12 mm = 85 mm

To work out the area of one brick, multiply these two dimensions:

234 mm × 85 mm = 19 890 mm²

In order to find out how many bricks you can get into 1 m², divide the area of one brick into 1 m². Before you can do this, you need to convert the area of the bricks (19 890 mm²) into m²:

$$\frac{19\ 890\ \text{mm}^2}{10^6} = 0.019\ 89\ \text{m}^2 \text{ or } 19\ 890\ \text{mm}^2 \times 10^6 = 0.019\ 89\ \text{m}^2$$

Now you have all the units in m², divide 1 m² by the area of 1 brick in m² to give you the number of bricks:

$$\frac{1\ \text{m}^2}{0.019\ 89\ \text{m}^2} = 50.28 \text{ bricks} + 10\ \%\text{ wastage}$$

In practice we usually allow 55 bricks/m² for half brick walls and 110 bricks/m² for one brick walls. Remember, when estimating the quantity of bricks required, you must make allowance for breakages (during offloading and construction) as well as off-cuts. Usually this figure is accepted as 10%. It is always better to estimate slightly more than to have too little, which could bring the work to a stop. A particular brick could even be either out of stock or out of production.

You would usually require 0.60 m³ of mortar per 1 000 bricks and, because of bulking and waste, 1.0 m³ of sand for every 0.60 m³ of mortar. Order 1 m³ of sand per 1 000 bricks.

Cement

Normally a mortar ratio of 1:6 is used, i.e. 1 part cement to 6 parts sand (refer back to your notes on concrete). Therefore 167 litres of cement are needed for 1 000 litres (1 m³) of loose sand:

$$\frac{1\ 000\ \ell \text{ sand}}{6} = 167\ \ell \text{ cement}$$

A bag of cement is approximately equal to 33 litres and therefore using the ratio above would amount to 5 bags of cement to 1 m³ of sand:

$$\frac{167\,\ell}{33} = 5 \text{ bags}$$

In summary, if 0.60 m³ mortar is sufficient for 1 000 bricks (allowing for wastage) then 1 000 bricks need 5 bags of cement.

Table 4.1

Quantities	1 000 bricks
Cement	5 bags
Sand	1 m³
Water	?
Mortar	0.6 m³
Area of wall	18 m²

4.2.14 Cavity walls

Definitions

A wall constructed in two layers or skins with a space between them is called a cavity wall and is the most common form of external wall used for domestic building today. Usually the cavity is 40–60 mm wide.

In order to strengthen these cavity walls, metal ties called tie wires are inserted at regular intervals in the wall, thereby connecting them. Tie wires are usually inserted at every fourth brick course and are usually 700 mm apart.

Activity 2

You are working on a construction site where the site foreman asks you to carry out the following task:

Determine by means of calculations the quantities of material (brick, mortar, sand and cement) required to complete the remaining two walls of a lecture theatre. Go to your local hardware store and enquire about prices for these materials. Then calculate the cost for the supply of these materials. Compare your answers with other members of your class.

- Wall 1 is 60 m long and 3.5 m high. There are no doors and windows in this wall.
- Wall 2 has the same dimensions except that this wall has eight windows (2.0 m long × 0.9 m high each) and one door 2.5 m wide × 1.8 m high.

■ Both walls are cavity walls 280 mm thick. Standard brick sizes are used. Joints are 12 mm apart.

Hints:
1. Draw each wall in section.
2. Start by calculating the area(s) of the wall and the window and door openings.
3. Subtract these openings from the area of wall 2.
4. Calculate the area of the brick and divide into each wall to determine the quantity of bricks required. Remember to double up the quantity as you are dealing with a cavity wall and add extra for breakages.
5. Calculate your mortar, sand and cement requirements.

Activity 3

Draw an isometric view of a cavity wall with the following dimensions:
■ 4.5 m long × 1.8 m high
■ The thickness of the wall is 280 mm inclusive of a 60 mm wide cavity opening
■ Clearly show all dimensions.

Can you think of a reason why walls would be built in this manner? Is it to save space or cost, to make the house seem bigger, to make the room smaller or to give the builder some extra work?

It is none of these. The most important reason for constructing the walls in this manner is to prevent the inner wall from getting wet. Imagine what could happen if the inner wall got wet – you would have paint peeling off the wall, mould would set in, it would be cold in winter and present a health risk. It is therefore important that the cavity is not bridged in any way, as this will provide a passage for the movement of moisture. The cavity must remain clean and free from mortar.

Have you noticed that in modern houses you will find an air vent construction into the wall?

Sometimes these vents, called air bricks, are located both near the top and the bottom of the wall. The reason for this is to allow air to circulate through the cavity and in that manner dry out any moisture that may have penetrated inside. Look around you and see if you can identify these air bricks and their locations in the wall.

Some builders dispute the necessity of inserting air bricks into cavity walls as this lowers the thermal and sound values of the wall. A good compromise is to insert air bricks into cavity walls only where the outside wall is exposed to very wet conditions.

When next you are on a construction site, notice that most builders fill the cavity below ground level with a weak concrete mixture in order to create a solid wall below the ground.

4.2.15 Glass bricks

One of the most recent innovations in the building industry is the use of glass bricks or blocks for aesthetic purposes. However, they are not solely for aesthetic purposes as they exhibit similar properties to ordinary bricks.

Glass bricks are hollow blocks consisting of two separate sections of pressed glass that are sealed together at high temperatures. They are used for windows, glass brick walls, staircase walls, parapet walls, lighting rows, etc.

Glass bricks are good insulation against cold, have good sound absorbing capabilities and excellent resistance to pressure. They can be used almost anywhere that extra natural light is required and they have good load-bearing characteristics.

Glass bricks are either 240 × 240 × 80 mm or 190 × 190 × 80 mm in size.

During construction, glass bricks must not be clamped or forced into the masonry, but must be separated from the adjoining masonry by expansion and contraction joints. They are usually constructed with 10 mm joints and laid onto a dryish mortar mix as they do not absorb any water.

Figure 4.19 Various types of glass bricks

Activity 4

Go to a housing building site and have a careful look at the things listed here. Take notes and compare yours with those in this book.

- Type of brick
- Any defects
- Type of bond used
- Type of wall construction (single, double, cavity, etc.)
- Are the cavities clean?
- Size of the bricks used.

Measure an area of one square metre on a wall, mark it off with coloured chalk, then count how many bricks are within that designated area. Verify this with the calculation using the size of the brick (see section 4.2.13).

Self-evaluation 4.1

1. Complete the sentences:
   a. The earliest bricks were made from _____.
   b. In the moulding process, indentations such as _____ may be formed.
   c. _____ are a very old method used for stock and handmade bricks.
   d. _____ are badly cracked and mis-shaped bricks, produced when rain falls on a hot brick.
   e. _____ is a complete layer of bricks plus its mortar bedding joint.
   f. A wall constructed in two layers with a space between them is called a _____ _____.
2. State whether the following statements are **true** or **false**:
   a. The presence of iron oxide affects the strength of the bricks.
   b. The dimensions of a standard clay brick are $222 \times 106 \times 85$ mm.

    c. The temperature obtained during the firing stage can be as high as 1 200 °C.

    d. Black core is caused by excess carbon matter in the clay.

    e. Brick gauge is the height of a brick course.

    f. An English bond consists of alternate courses of headers and stretchers.

3. Answer the following:

    a. Explain why heat and humidity must be carefully controlled in a brick drying kiln.

    b. Why is it necessary to have standard brick sizes?

    c. Why do you think it is necessary to perform tests on bricks?

    d. When using engineering bricks, do you think it is important to specify the strength requirement and why?

4.3 Timber

Timber is one of the oldest structural materials. Temples and monuments built several centuries ago that still exist today demonstrate the durability and usefulness of timber. The advent of other building materials, such as steel, bricks and concrete, revised building techniques, the growing shortage of wood resources and the increased costs of timber structures have resulted in the decline of the use of timber in major construction.

South Africa's main forestry areas, providing species such as teak, yellow wood, black wood, stink wood and pine, are in KwaZulu-Natal, Northern Province, Mpumalanga, and Eastern and Western Cape. Hardwoods such as beech, oak, teak and embuia are all imported into South Africa. Oak is especially important as it is used in the wine industry to make casks for the storage and maturation of wines and spirits.

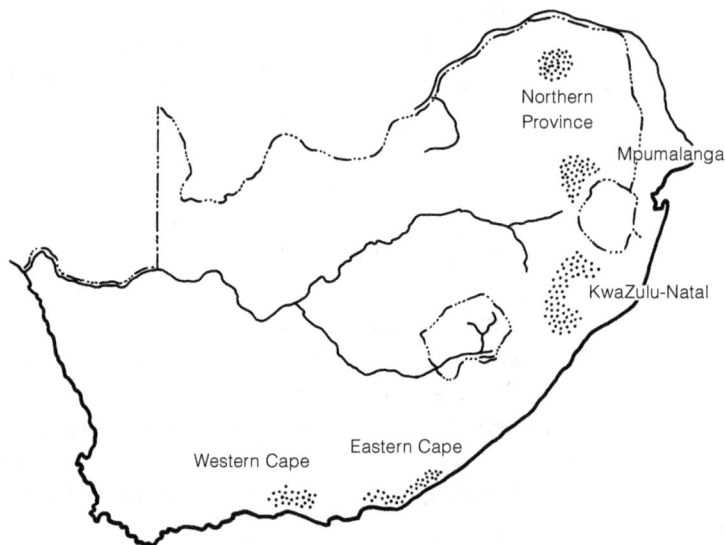

Figure 4.20 Map of South Africa showing where wood resources are

In order to enhance and economise the utilisation of wood, many woodbased products have been developed, for example veneers, plywood, hard board, press board, etc. These have found wide acceptability in the market for various uses from furniture making to partitions, interior decoration and, of course, as formwork for construction purposes.

4.3.1 Anatomy of a tree

It is important to know something about the nature of trees to get a good understanding of the properties, classification and use of timber.

A tree is basically made up of roots, a stem, branches and leaves. Underneath the bark of a tree is a delicate tissue known as **cambium.** This forms a complete sheath around the stem and the branches. Cambium is responsible for the growth of a tree and produces bark towards the outside and wood towards the inside. The cambium is active only during the growing season. Growth occurs by continuous division of cells. Growth is usually limited to one season during the year.

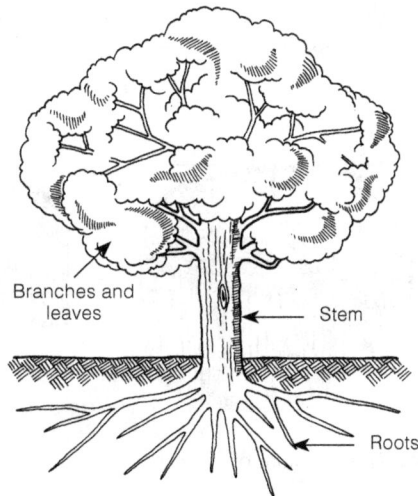

Figure 4.21 A tree and its parts

The trunk also grows in length in what is called **terminal growth.** Since the tree normally grows only during one season in a year, the tree adds one ring to its girth each year. The growth can normally be divided into two parts, early growth and late growth. In many trees there is a marked difference between these two and this gives rise to what are known as annual rings. The number of rings normally indicates the age of the tree, with one ring added each year, but there are exceptions.

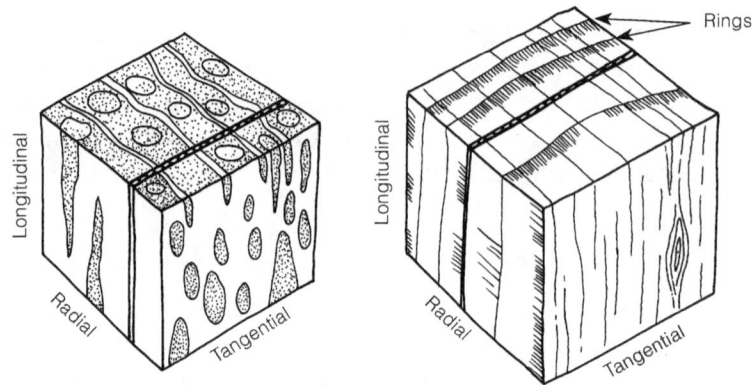

Figure 4.22 Three directions of timber growth

Fig 4.22 shows three ways in which the direction of growth rings can relate to surfaces. A line down the length of a log represents the **longitudinal direction** in the log. A line drawn from the centre of the log to the bark, i.e. cutting across the annual growth rings, represents the **radial direction**. Any line perpendicular to both the radial and the longitudinal direction, i.e parallel to the annual rings as seen on a cross-cut stem, is called the **tangential direction**.

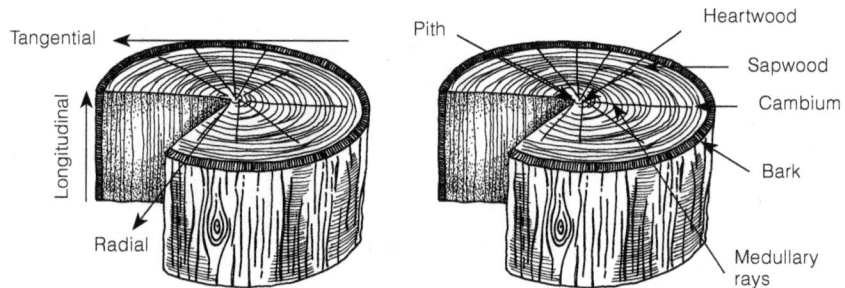

Figure 4.23 The structure of wood

The central part of the cross-section area of the stem of a tree is usually considerably darker than the ring of wood next to the bark of a tree. This is more distinct in hardwoods than in softwoods. This inner, darker part of the stem is called the **heartwood** while the outer, lighter coloured wood is called **sapwood**.

The proportion of heartwood to sapwood in a tree depends on growing conditions and species. Where it is difficult to distinguish between sapwood and hardwood, the freshly cut surface can be treated with a chemical stain to show up the heartwood.

4.3.2 Moisture content

Trees also contain moisture and, like soils, this can have an effect on volume and density. You can calculate the moisture content of timber in a similar manner to that of soils. Usually wood density is quoted at either no moisture or a specific moisture content of 12%.

The density of wood varies of course, dependent on factors like moisture, species and cell structure.

Activity 5

Find a piece of freshly cut wood. Cut three equal lengths from it. Measure the weight of all three pieces on a scale. Record these masses. Place the first piece in the sun and leave it to dry for five days. Weigh and record its mass. Put the second piece in an oven for about 16 hours at low heat. Weigh and record its mass, Place the third piece in a container with water and leave it submerged for 24 hours. Weigh and record its mass.

Now compare the three masses and calculate their moisture contents. You will see how each has a different moisture content and notice the effects of the different environmental conditions on each. Lastly, have a look at the change in the appearance of the wood, if any, and discuss this with your classmates.

4.3.3 Classification of trees

Trees can be classified into two groups:
- **Exogenous** (exogens), which grow outward by the addition of a new sheath of wood every year, e.g. teak
- **Endogenous** (endogens), which grow inward from a hard exterior shell or, more commonly, endwise by the acquisition of new joints, e.g. bamboo and palm trees.

Another important classification is **hardwoods** and **softwoods**. The classification is no reflection on how hard or soft the wood is. There are many softwoods that are a lot heavier or denser than some hardwood species. The hardwoods are broad-leaved trees and have entirely different characteristics to softwoods, which have needle-shaped leaves and grow in temperate regions and high altitudes. Examples of hardwoods are meranti, embuia, oak and stinkwood; softwoods include pine and spruce.

> *The classification of a tree as a hardwood or softwood has nothing to do with how hard or soft the wood is.*

Softwood trees have mainly one type of cell called **tracheids** which convey sap and support the tree. Hardwood trees contain two kinds of cells – **vessels** and **woodfibres**. Vessels are large, thin-walled cells that convey sap solutions from the root to the leaves. Wood fibres are small, thick-walled cells that serve to strengthen and support the tree. These cells are bound together by lignin.

4.3.4 Comparing hardwoods and softwoods

Table 4.2 lists the differences between hardwoods and softwoods.

Table 4.2

Hardwood	Softwood
Deciduous – lose leaves in autumn	Coniferous – usually evergreen
Angiosperm – enclosed seeds	Gymnosperm – naked seeds
Leaves broad and flat	Leaves are spiky and needle-like
Branches usually grow out at different levels	More than two branches at the same level
Cells are vessels and fibres	Cells are tracheids
Trees are short and squat	Trees are long, tall and comparatively thin
Found throughout the world	Found in temperate regions, not tropical
Growth rings not as distinct	Growth rings distinct
Usually heavier in weight	Usually light in weight
Generally dark coloured	Light coloured
Non-resinous and do not catch fire readily	Resinous in nature and readily catch fire
Cannot be split easily and are strong in tension, compression and shear, strong along and across the grains	Can be easily split and are strong along the grain but weak across it

4.3.5 Other methods of classifying timber

Based on modulus of elasticity: The species of timber recommended for constructional purposes are classified into three groups:
- Group A: modulus of elasticity in bending above 125 t/cm^2

- Group B: modulus of elasticity in bending between 98 t/cm^2 and 125 t/cm^2
- Group C: modulus of elasticity in bending above 56 t/cm^2 and below 98 t/cm^2.

Any timber below 56 t/cm^2 is unacceptable for structural timber.

Grading of structural timber: Structural timber can be graded into three classes, namely select grade, grade I and grade II. This classification is based upon the structural design characteristics like permissible stresses, defects, etc.

Based on availability: Timber can be of three categories, classified by availability within South Africa:
- X – most common: 1 415 m^3 or more per year
- Y – common: 355–1415 m^3 per year
- Z – less common: below 355 m^3 per year.

Durability: Test specimens of 60 × 5 × 5 cm are buried in the ground to half their lengths by the forestry department and the conditions of the specimen at various intervals of time are noted and their average life calculated. These are then classified as having:
- High durability: an average life of 120 months and longer
- Moderate durability: an average life of 60–120 months
- Low durability: average life of less than 60 months.

Seasoning characteristics: Timber can be classified depending on the way it cracks and splits during normal air-seasoning:
- Highly refractory (class A) are slow and difficult to season without defects
- Moderately refractory (class B) can be seasoned free from surface defects
- Non-refractory (class C) can be rapidly seasoned free from defects.

Treatability: The behaviour of timber to preservative treatment under pressure can be classified as:
- Easily treatable
- Treatable but complete preservation is not easily obtained
- Only partially treatable
- Refractory to treatment
- Very refractory to treatment, penetration of preservative being practically nil from both sides and ends.

4.3.6 Identification of timber

How can timber be identified?

There are seven ways timber can be identified:

- **Colour:** When you walk into any timber store you will notice the different colours or shades of timber. These can range from creamy white to jet black, through varying shades of grey, yellow, pink, brown and purple. Sapwood is generally lighter than heartwood. Darker colours in wood indicate greater durability because of the presence of natural toxic substances
- **Odour:** Most woods do not have any characteristic odour that differentiates them from others. Their odour disappears on exposure and, if present, is pronounced only when freshly cut
- **Hardness:** This is the resistance to identation or penetration by a foreign body and does not strictly apply to timber. However, if you test the hardness of different timbers by pushing a fingernail or pen point into them, you will find that some timbers are easier to indent than others
- **Density:** This property of wood is of great significance to the timber user. From a structural point of view, the higher the density, the stronger the timber. This is because a dense timber has a more compacted structure. Density values are determined at a specific moisture content of 12%
- **Grain:** This is the general direction or alignment of the wood cells. The nature of the grain affects the stength, seasoning and other properties of the timber. Grain that is not straight is seen as a defect in timber
- **Texture:** The size of the wood cells, their distribution and the proportion of the various types of wood cells all affect texture. Remember that grain refers to the alignment of the wood cells; the way a timber feels to the touch is due to its texture. Coarse-textured timber will be rough to the touch, whereas fine-textured timber is smooth
- **Lustre:** The way light reflects from the wood cells is known as lustre. A piece of timber can contain cells that reflect light differently, e.g. the silver grain in oak.

4.3.7 Seasoning

Seasoning is the drying of the sap and the moisture in all the cell cavities and cell walls, thus reducing the moisture content of the wood to that of the atmosphere where it is to be used. Shrinkage is reduced to a minimum. There are two types of seasoning: natural and artificial.

Natural seasoning

When seasoned naturally, timber is stacked either out of doors where the piles are protected from rain by temporary sloping roofs or in an open shed having a roof and one or more walls. The site or floor should be well drained and covered with ash or, preferably, concrete to prevent the growth of vegetation. The standard width of a pile of timber varies from 1.8–3.7 m, the height may be up to 5 m and the length depends upon that of the timber. Piles are best built on steel beams or rails supported at intervals by 228 mm^2 concrete or brick piers. If piers are not used, the timber in contact with the floor must be creosoted.

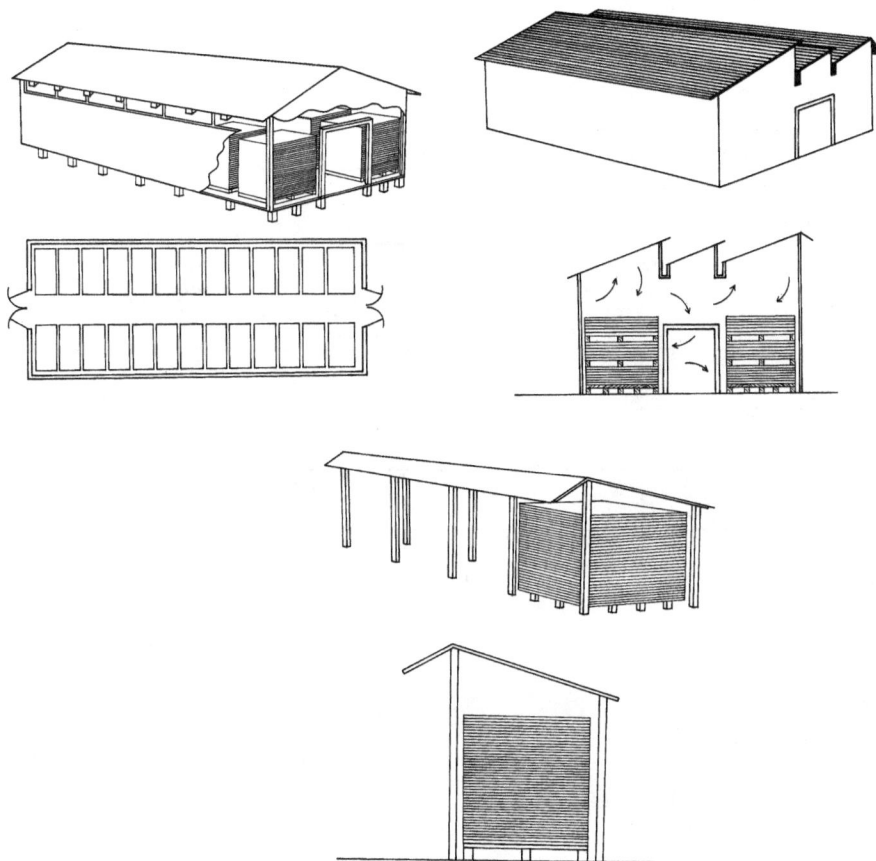

Figure 4.24 Examples of sheds and coverings

Softwoods in the form of baulks are arranged so that there is a free circulation of air between and among the piles thus reducing the moisture content of the wood. The time taken to dry out depends on the temperature and humidity of the atmosphere, the thickness and density of the timber and the efficient piling or stacking. On average it takes about 2–3 months to reduce the moisture content in 25 mm thick softwood boards to 20% (the acceptable moisture content) and 50 mm thick pieces will take 3–4 months. Hardwoods take longer. Table 4.3 shows the advantages and disadvantages of natural seasoning.

Artificial seasoning (kiln seasoning)

Kiln seasoning is used on a vast scale as it provides controlled conditions for rapidly drying out timber to any required moisture content. The timber is stacked in a kiln of which there are several types. Heated air is circulated through the timber. This hot air accelerates the evaporation of moisture from the timber. It must contain a certain amount of moisture or else the wood will dry too rapidly, causing the splitting and hardening of the wood. Splitting happens when the outside of the wood dries faster than the interior.

Table 4.3

Advantages	Disadvantages
Relatively cheap for small supplies	Rate of drying is very slow
Requires little attention	Cannot be rigidly controlled
Defects due to the process are generally small	Even under favourable conditions, the moisture content cannot be reduced to that required for certain internal joinery
	Large stacks of timber require considerable space
	Much capital is unproductive for a lengthy period
	Damage to timber may be caused by fungi and insects

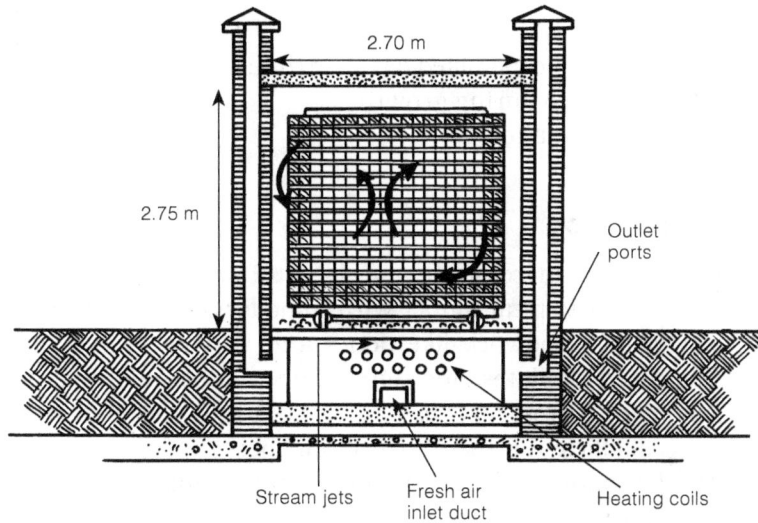

Figure 4.25 Artificial seasoning done in a kiln

4.3.8 Conversion (sawing)

There are various **ways of converting** a log into planks, boards, etc:

- Radial, rift or quarter sawing
- Tangential sawing
- Slab sawing.

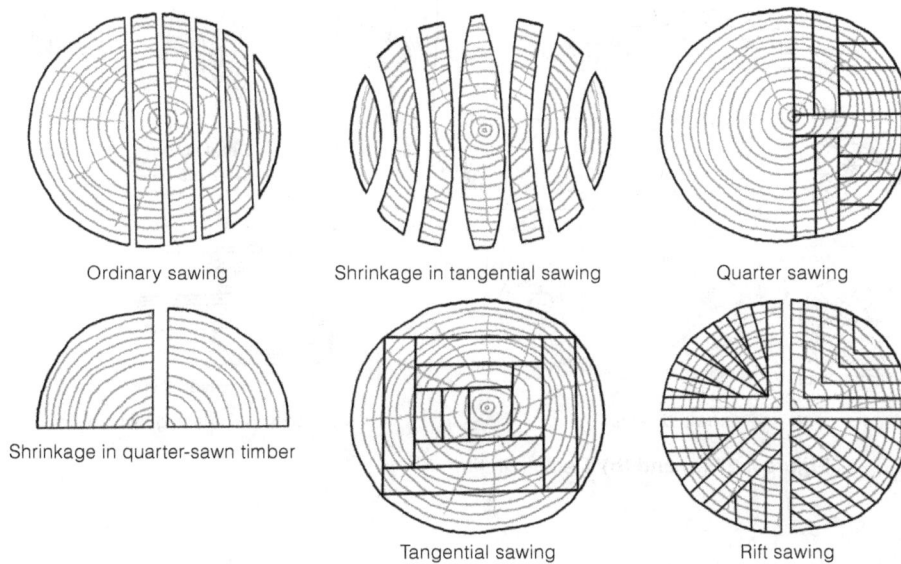

Figure 4.26 Various ways to convert a log

4.3.9 Defects

Defects can be classified as those that:
- Developed during growth
- Occurred after felling.

4.3.10 Preservation

In order to increase the durability of seasoned timber, it is sometimes necessary to treat timber with some form of preservative. For example, external woodwork like doors and window frames are painted and fences, wall plates, floor joints and weather boarding are protected by applying a preservative.

An efficient preservative should be:
- Poisonous to fungi and not human beings
- Have good penetration of timber
- Be cheap and readily available
- Should not corrode metal
- Should not render the material more flammable.

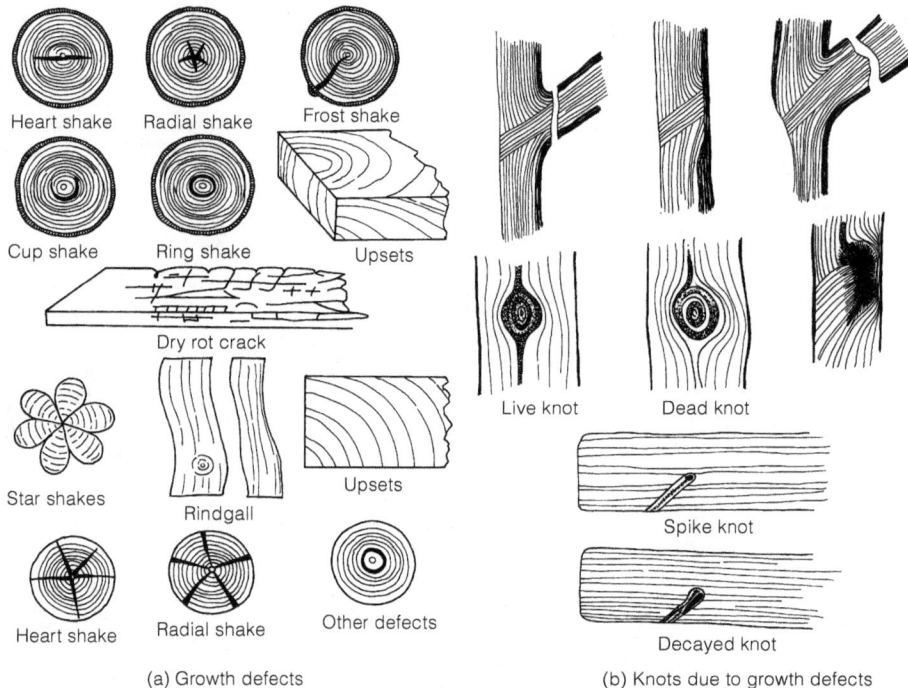

(a) Growth defects

(b) Knots due to growth defects

Figure 4.27 (a) and (b) Defects in timber

⇨

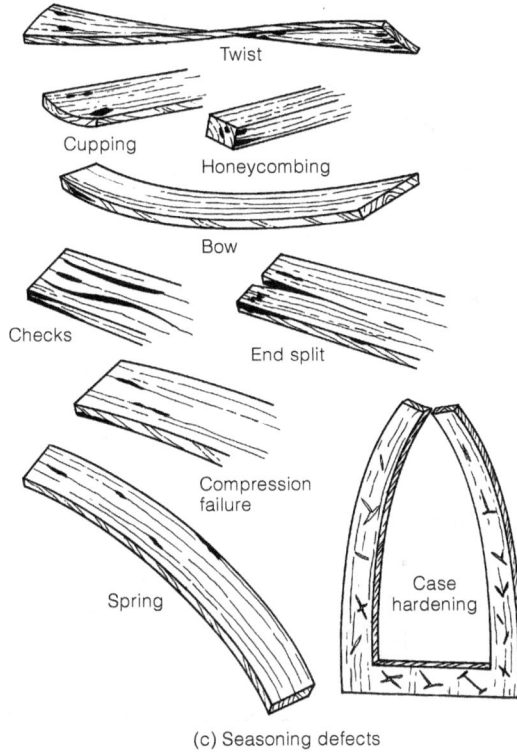

(c) Seasoning defects

Figure 4.27 (c) Defects in timber

Types of preservatives

Can you name three types of preservatives?

Creosote is an effective general purpose preservative, particularly for external work. Creosote-treated timber cannot be painted and so it is unsuitable for exposed timber work in internal positions. It is a black or brownish oil and is produced by the distillation of coal-tar, a by-product in the manufacture of gas. It is poisonous to fungi and insects. It has a strong smell. It is permanent when properly applied, cheap and readily

221

available. Creosote is widely used in the treatment of railway sleepers, fences, telephone poles and gates.

Organic solvents are various toxic chemicals in an oil base, white spirits or petroleum distillates. They cost more, have good penetration and dry quickly, for example pentachlorophenol.

Water-based metallic salt preservatives are mostly based on salts of copper, zinc, mercury, sodium and chromium dissolved in water. They are resistant to leaching (cannot be washed away), do not evaporate quickly and retain their strong poisons to prevent fungal decay and the activities of wood destroying insects. Timber can be treated, dried and then painted or varnished. It is odourless, non-oily and non-staining so treated timber can be used internally and for food storage without risk of contamination.

4.3.11 Methods of applying preservatives

Preservatives can be applied by pressure processes, non-pressure processes and superficial processes.

Pressure methods are generally adopted for treating timber on a large scale and for maximum penetration. They include full cell processes and empty cell processes. Pressure processes use a steel cylinder about 7 m in length and 1.5–3 m in diameter to treat timber. The timber is loaded in trolleys and then wheeled into the cylinder.

The cylinder is closed and a vacuum created, whereafter the preservatives are introduced until the cylinder is completely filled. The timber is left in the tank to allow the preservative to 'soak' into it. The preservative is then drained off and the timber removed from the tank. The time the timber is left in the tank depends on the wood species and size of the logs.

With the **full cell process**, a vacuum is first created within the timber, after which it is placed into the tank of preservative and only then is the pressure applied.

The **empty cell process** involves compressing the timber with air pressure before introducing the hot preservative and applying even greater pressure until an excess absorption of preservative is secured.

Storage tanks for the preservatives with pumps, gauges, steam pipes, etc. are also required.

The application of preservatives by **non-pressure methods** is generally less effective than that by pressure methods. Brushing and spraying are principally used for on-site application of solvent preservatives and light-oil creosote. These types of application are used mainly for maintenance and remedial treatments and are adequate for low fungal decay situations. A two- or three-coat treatment is usually sufficient for timber not in contact with the ground.

Superficial methods include:

- **Dipping,** which is the best method of surface treatment, except for timber already in position. The pieces of wood are just dipped in a container holding the preservative. The longer the immersion, the better
- **Spraying** applies the preservative in a fine spray as it is forced through the nozzle of the appliers by compressed air. It is an effective form of surface treatment as the pressure makes the penetration of any cracks or crevices possible
- **Application by brush** is the most common method of treating exposed woodwork with creosote or other preservative. The liquid should be applied liberally with a brush. At least two good coats should be given, the first being allowed to dry before the second is brushed on. Where possible, this treatment should be renewed every three years, especially if it is for external work such as gates, fencing and timber outbuildings.

The choice of treatment is governed by the timber type and moisture content as well as the use to which it will be put.

> What influences the choice of treatment?

Fire-retardant timber
The combustion of timber can be delayed by treating the wood with chemicals. These produce non-flammable gases that reduce the availability of oxygen to the interior of the wood.

4.3.12 Wood-based products

In order to economise on the use of timber, which is a natural resource that is in limited supply, many wood-based products like veneers and plywood have been developed.

Veneers
Veneers are thin sheets of wood, 0.4–6 mm thick, obtained by different

knife cutting processes. Depending on the cutting process, veneers can be classified as

■ Rotary veneers
■ Sliced veneers.

Rotary cutting method

More than 90% of veneers are cut by this method. A rotary veneer cutter or peeler is a powerful lathe with a very sharp fixed knife slightly longer than the log. The log is conveyed by a crane to the peeler, lowered and then clamped between two centres which penetrate the ends of the timber. The horizontal log is revolved and the knife cuts a continuous ribbon of veneer, of uniform thickness. What emerges is like a roll of paper being unrolled. A pressure bar prevents the wood from splitting. The distance between the bar and the knife is regulated according to the thickness of the veneer required. Logs converted in this manner should be free from knots and other defects and be reasonably straight grained.

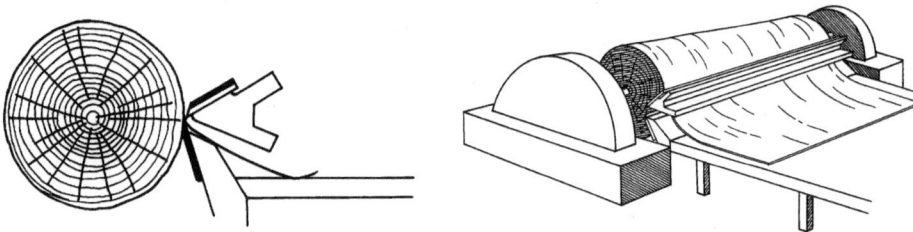

Figure 4.28 Rotary cutting method

Veneer slicing method

Decorative veneers are obtained from certain rare, richly figured timbers by slicing so that the attractive figure may be shown to greater advantage than that produced by the rotary cutter.

Plywood

Plywood is a compound wood made of several thin layers of plys or veneers glued together under pressure and usually arranged so that the grain of one layer is at right angles to the grain of the adjacent layer(s).

Plywood is obtainable in many kinds of wood and there are different grades offering varying resistance to weather; some are classed as weatherproof – others are only suitable for internal use.

A sheet or board of plywood usually consists of an odd number of plys, i.e. 3-ply, 5-ply, etc. Those which have more than three layers are known as multi-ply boards. A 3-ply board consists of two outer or face plys with a middle core. It is important to observe that these plys are cross grained, i.e. the grain of the core of a 3-ply board is at right angles to that of each of the face plys. The thickness of the veneers varies from 1.6–16 mm.

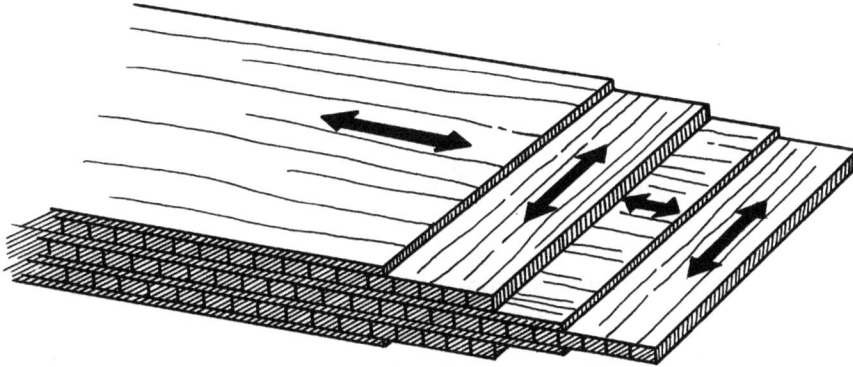

Figure 4.29 Assembly of plywood layers

Self-evaluation 4.2

1. Complete the sentences:
   a. Wood used for structural purposes is known as_____ .
   b. _____ is a term used to describe the support given to concrete when wet.
   c. The inner, darker part of the stem is called the _____ .
   d. _____ is the drying up of the sap and moisture in all the cell cavities.
   e. _____ is an effective general purpose preservative particularly for external work.
   f. _____ is a compound wood made of several thin layers of plys and veneers.
2. State whether the following statements are **true** or **false**:
   a. Wood is a natural resource.
   b. The cambium produces bark towards the inside and wood towards the outside.
   c. Exogens grow outward and endogens inward.
   d. Vessels convey sap solutions from the root to the leaves.
   e. Softwood leaves are broad and flat.
   f. Brushing is a method of applying creosote or other forms of preservative.
3. Answer the following:
   a. Name the parts of a tree and describe their functions.

b.   Describe the classification of trees.

c.   Describe the differences between natural and artificial seasoning methods.

d.   What is a preservative and what properties must it exhibit?

e.   How is plywood obtained?

f.   What function do the tracheids in softwood perform?

4.4 Metal

Metal is a broad term used to describe a variety of material, for example aluminium, steel, copper, gold, lead, zinc, tin and silver. Generally metal is regarded as a tough, hard and durable material able to withstand high tensile and compressive forces, i.e. it is very strong.

Metals are grouped into two main categories: **ferrous** and **non-ferrous**. The difference between the two is that ferrous metals are made predominantly from iron whereas non-ferrous metals are not.

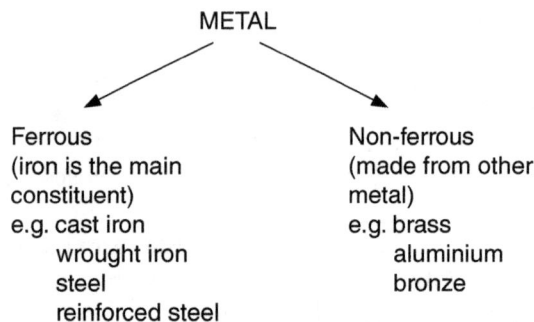

METAL

Ferrous
(iron is the main
constituent)
e.g. cast iron
    wrought iron
    steel
    reinforced steel

Non-ferrous
(made from other
metal)
e.g. brass
    aluminium
    bronze

Figure 4.30 Ferrous and non-ferrous metals

We will be concentrating mostly on the ferrous metals and, in particular, steel. First you must familiarise yourself with several definitions:

Definitions

Stress: The force in Newtons applied to a section (whether a beam, column, etc.) divided by the area of that section where the force is applied. It is usually expressed in N/mm^2 or MPa.

$$\text{Stress} = \frac{\text{force}}{\text{area}} = \frac{\text{force applied (N)}}{\text{area of section (mm}^2)}$$
$$= N/mm^2$$

EXAMPLE 1

A bar with a cross-sectional area of 10 mm × 10 mm is loaded with a force of 1 000 N. Determine what the stress at the point of loading is.

$$\text{Stress} = F/A = \frac{1\ 000\ N}{100\ mm^2}$$
$$= 10\ N/mm^2$$

Figure 4.31

Strain: When a force is applied to a section and that section changes in length, the change is expressed as a ratio. For example, take an elastic band of a known length, hold the one end and pull it from the other end. What happens? The elastic band extends. To determine the strain experienced in the elastic band, you have to calculate the change in length and express it as a ratio. In other words:

$$\text{Strain} = \frac{\text{change in length}}{\text{orginal length}}$$

Note that the change in length can either be an increase or a decrease. Take, for instance, a normal eraser and squeeze from both sides. The length of the eraser will decrease also resulting in a strain.

EXAMPLE 2

Apply a force of 1 000 N to a steel reinforcing bar by fixing the bar on the one end and pulling from the other. If the bar is initially 1 000 mm long and after application of the force it is now 1 050 mm, what would the strain experienced in the reinforcing bar be?

$$\text{Strain} = \frac{\text{change in length}}{\text{original length}} = \frac{(1\ 050 - 1\ 000\ mm)}{1\ 000\ mm}$$
$$= 0.05$$

There are two forms of strain: **elastic strain,** like the elastic band that will return to its original or near original shape, and **plastic strain (permanent deformation)**, like the steel reinforcing bar that will never return to its original shape. (Also refer back to the description of plastic deformation on page 98.)

Elastic limit: The stress beyond which any increase in load causes permanent deformation.

Yield: The permanent deformation a steel piece takes when it is stressed beyond the elastic limit.

Modulus of elasticity or **Young's modulus:** The ratio of the stress (F/A) to the strain (deformation) in the elastic range. It is expressed in units of stress, i.e. N/mm² or MPa.

$$\text{Young's modulus} = \frac{\text{stress}}{\text{strain}} = \frac{(\text{force/area}) \text{ N/mm}^2}{\text{deformation}}$$

Steel has a typical value for Young's modulus of 2.1×10^5 MPa.

Moment of inertia: The resistance to bending in a beam section. This is dependent on its shape and size

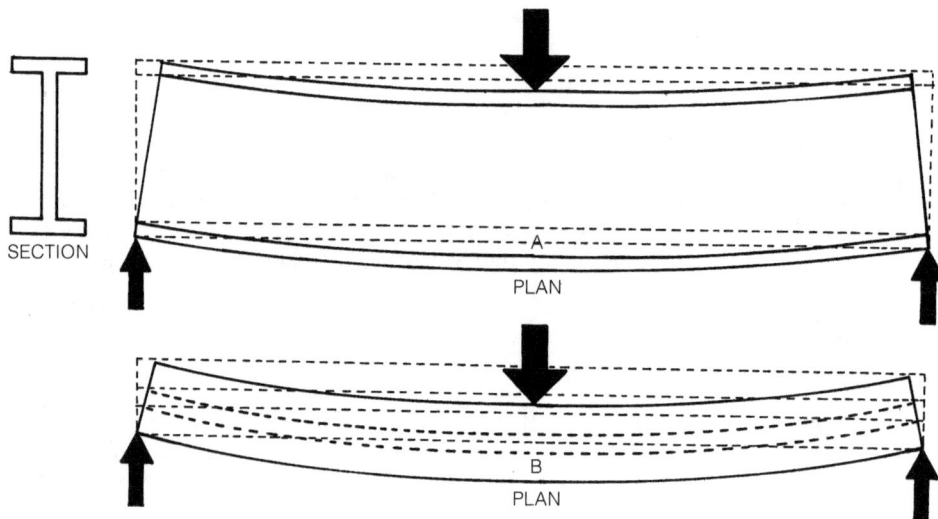

Figure 4.32 A steel beam in plan and section showing imposed bending. In position B, the steel beam is more apt to deflect than in position A.

Neutral axis: The line of zero stress in a bent beam. Below the neutral axis all fibres are stressed in tension and all fibres above this line are in compression.

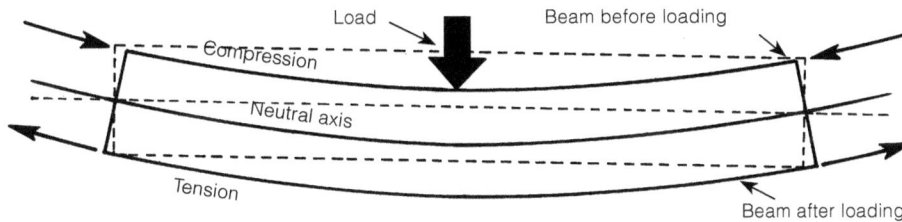

Figure 4.33 A steel beam in section showing the neutral axis running along the centre of the beam as well as indicating compression in the top and tension in the bottom

Compression: A force which tends to shorten a member. It can also be viewed as a pushing force like when you squeezed the eraser earlier.

Tension: A pulling force or stress which tends to lengthen the member. The elastic band is a good example of a pulling force. Steel is strong in tension.

4.4.1 Ferrous metal

Ferrous metal is divided into three broad categories: cast iron, wrought iron and steel. All three are obtained from pig iron and undergo certain processes. The main distinguishing factor is the amount of carbon that each contains. Cast iron contains the highest percentage of carbon whereas steel has the lowest. The higher the amount of carbon, the harder and stiffer the metal.

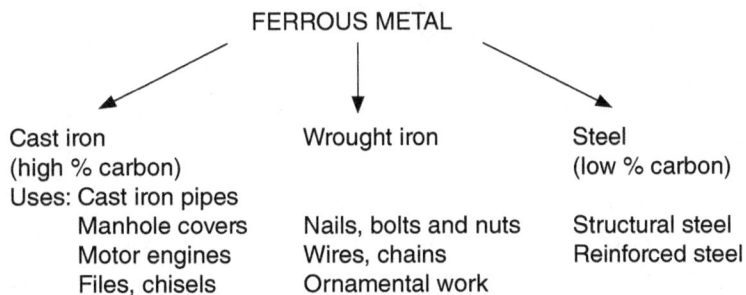

FERROUS METAL

Cast iron
(high % carbon)
Uses: Cast iron pipes
      Manhole covers
      Motor engines
      Files, chisels

Wrought iron

Nails, bolts and nuts
Wires, chains
Ornamental work

Steel
(low % carbon)

Structural steel
Reinforced steel

Figure 4.34 Ferrous metal

4.4.2 Steel

Steel is a mixture of materials usually iron, carbon, manganese and a very small percentage of other elements. The mixture of any two or more elements is known as an **alloy**. There are various methods of making steel, which have been refined with time. Initially, two methods of refining steel were used: the **Bessemer process** and the **open-hearth** process. Since 1970, these two methods have been replaced by the **basic oxygen process** and the **electric arc furnace process**.

4.4.3 Manufacture of steel

What are the properties of mild and high tensile steel?

Liquid pig iron is obtained by smelting the raw materials. The liquid pig iron is refined until the pig iron is purified to the desired extent, whereafter the molten steel is drawn from the furnace and poured into ingot moulds. When required, these ingots are rolled and shaped into a variety of forms, for example steel bars, structural steel members, wires, sheets and pipes.

In addition to the amount of carbon present, steel can further be divided into four categories, the most important being **mild steel** and **high tensile (carbon)** steel. The properties of each are given in table 4.4.

Table 4.4 Properties of steel

Mild steel	High tensile steel
It has a fibrous structure and appears dark bluish in colour	It has more of a granulated structure
It can be easily welded, riveted and forged	It is more difficult to forge and weld
It is equally strong in tension, compression and shear	It is stronger in compression than in tension or in shear
It is difficult to harden and temper	It is tougher than mild steel

4.4.4 Products made from steel

Have you noticed any of the steel shapes represented by the illustrations of sections in fig 4.35 when passing a construction site or looking at a photograph? Where and how do you think these structural steel items can be utilised? If you are thinking of steel beams and columns, you are on the right track.

Beam/column

Channel

Angle

I-section

Parallel flange

Equal leg

Hollow sections

Circular

Square

Reinforced steel

Mild steel rolled bars

High tensile steel rolled bars

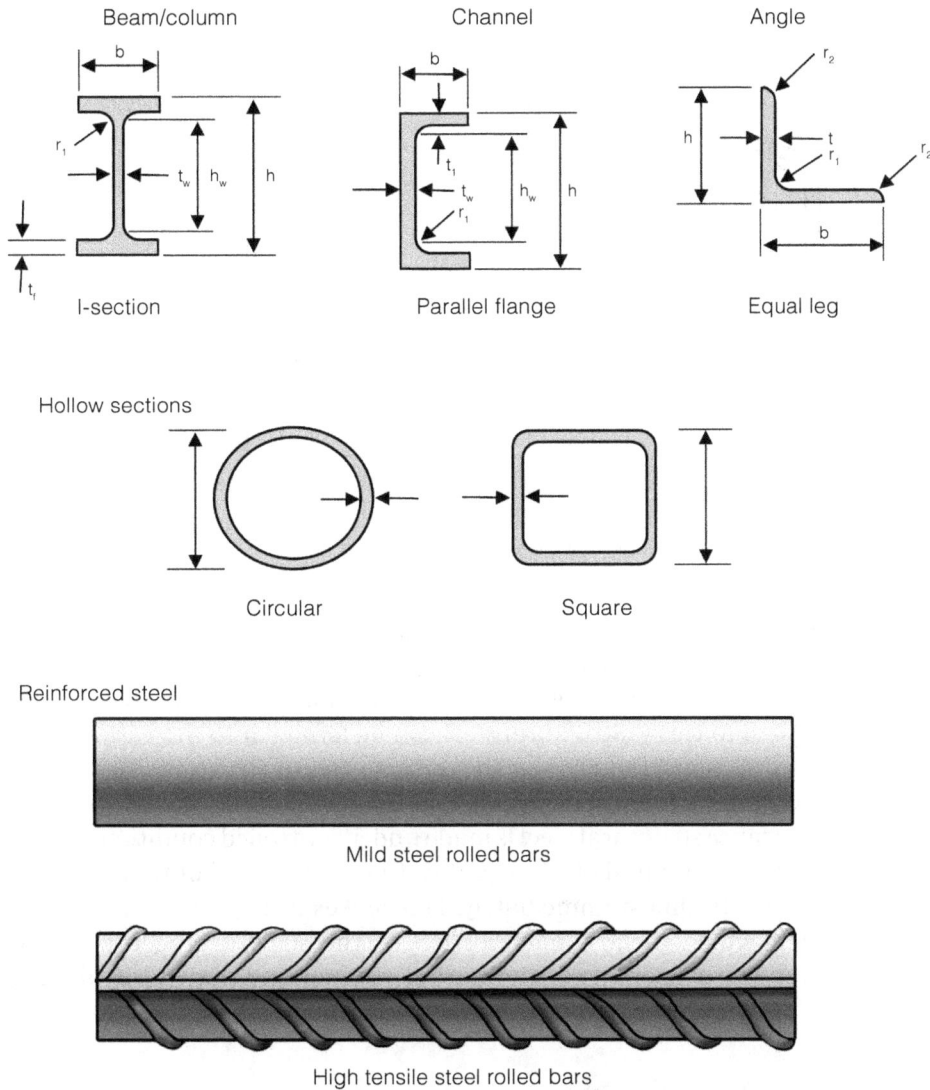

Figure 4.35 Structural steel

4.4.5 Structural steel

Industrial buildings are often designed and constructed using a steel structure. This structure is made from a combination of steel columns and beams as well as a roof truss made from steel. All these sections are either bolted, riveted or welded together.

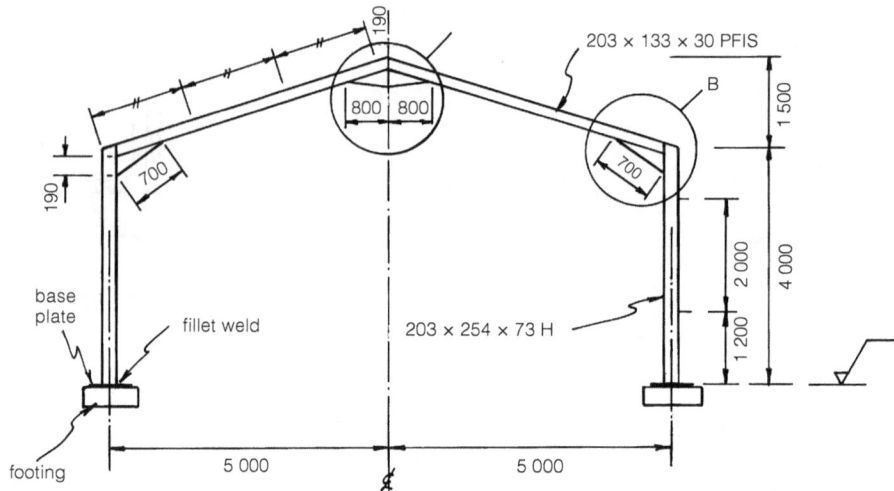

Figure 4.36 A steel structure with various sections

Structural steel is composed almost entirely of pig iron with small quantities of other elements added to give it strength and ductility. This strength makes it possible to withstand the large loads that it will be required to carry. Structural steel can be formed into the various shapes (called sections) required for the certain types of structures (see fig 4.36). You will learn more about this in *Construction Methods for Civil Engineering*.

Because structural steel is made under controlled conditions, the user is generally assured of high quality. Due to the speed of manufacture, it is available almost immediately. This makes transport of these sections faster, which means that erection of the structure occurs so much faster. The availability of the various sections and shapes has also made the design of steel structures easier and also less costly.

4.4.6 Structural steel sections

There are numerous shapes of sections all available in various sizes to suit most conditions encountered on site. Most of these sections are available in two grades, namely grade 43 for the lower strength steel and grade 50 for higher. A complete list of sections and their properties is in the South African Institute of Steel construction handbook called *Structural Steel Tables*.

Steel sections are specified in a standard manner and these notations must be adhered to, so as to avoid confusion. Imagine you are designing a member with a certain length or thickness but the contractor interprets your design differently and orders the wrong member. Steel members are specified by height × width × mass per metre, followed by the shape. When you read a specification on a drawing you will find, for example, the following: 203 × 133 × 25 kg H, meaning that it is an H-beam with the dimensions of 203 mm height × 133 mm width × 25 kg per metre.

There are three common sections that are specified on structural steel members: **H-sections**, **I-sections** and **T-sections**. H- and I-sections are normally used as beams, rafters, columns and portal frames. T-sections are most commonly used as the bottom or top chords in roof trusses. They are obtained by cutting I-sections in half.

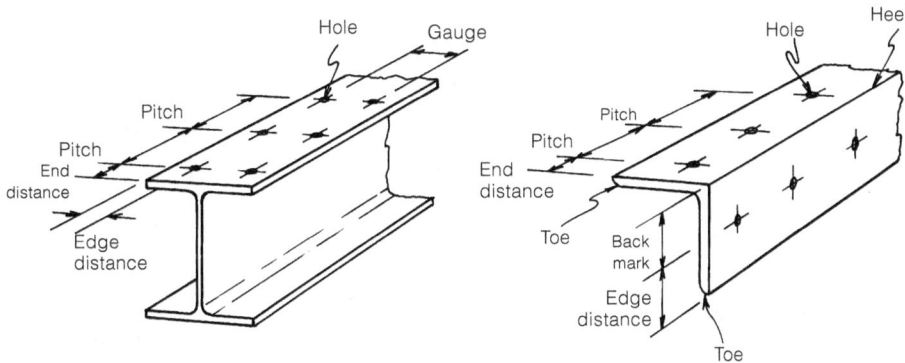

Figure 4.37 Parts of a section showing hole positioning terms

4.4.7 Reinforcement steel

Section 2.7 dealt with where and how steel reinforcement is used. There are various types of steel reinforcement available in South Africa that are covered by the South African Bureau of Standards (SANS) steel specification.

In your drawing classes at S2, you will come across methods of detailing reinforced concrete sections as well as the various shape codes, areas, spacings and lengths applicable. You will also be required to complete the bending schedule as is necessary when working with reinforced concrete. For your convenience and cross reference, fig 4.38 gives the shape codes.

SUMMARY OF SHAPE CODES

(Reference: SANS 282:2004, Edition 5.1 – Bending dimensions and scheduling of steel reinforcement for concrete)

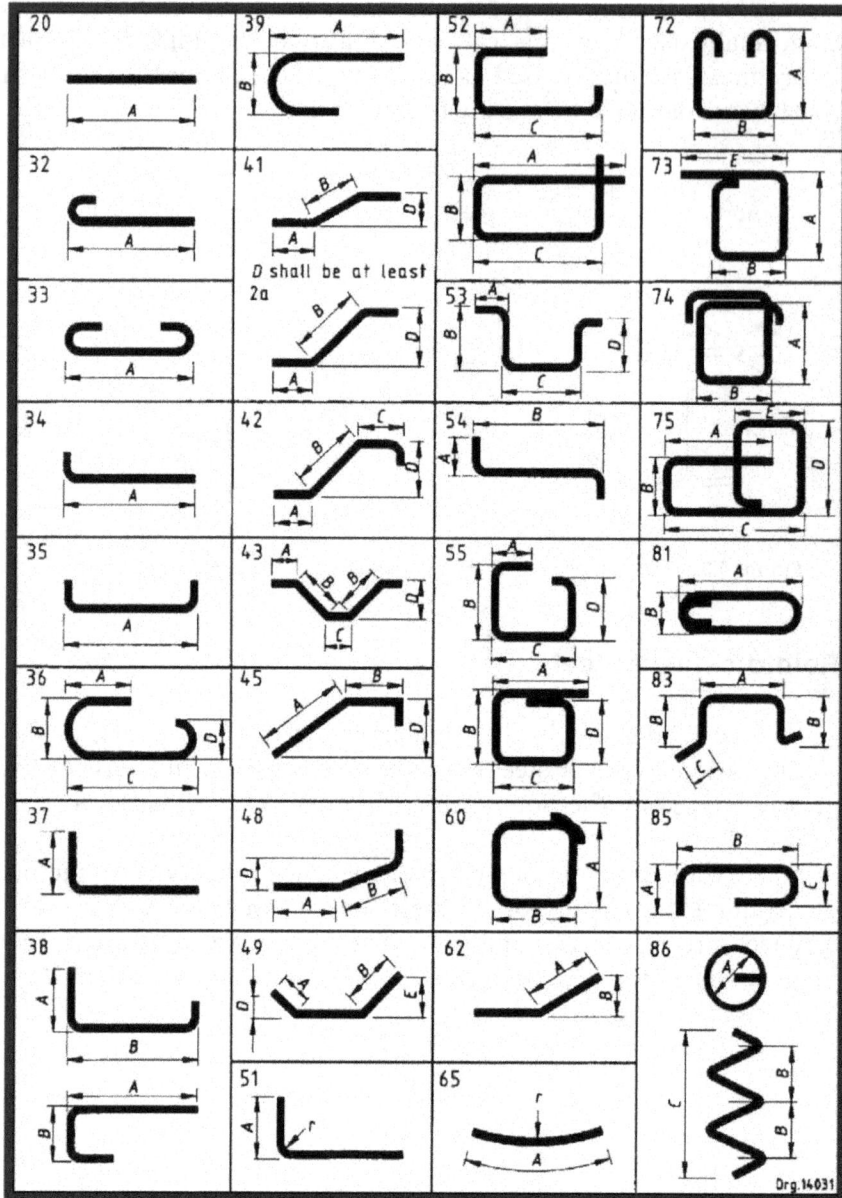

Figure 4.38 Summary of shape codes material reproduced from SANS 282: 2004

Reinforcement 1

The specified concrete cover should be maintained by an approved means, using either concrete or plastic spacers of the correct size fixed between the reinforcement and the shutter or blinding concrete.

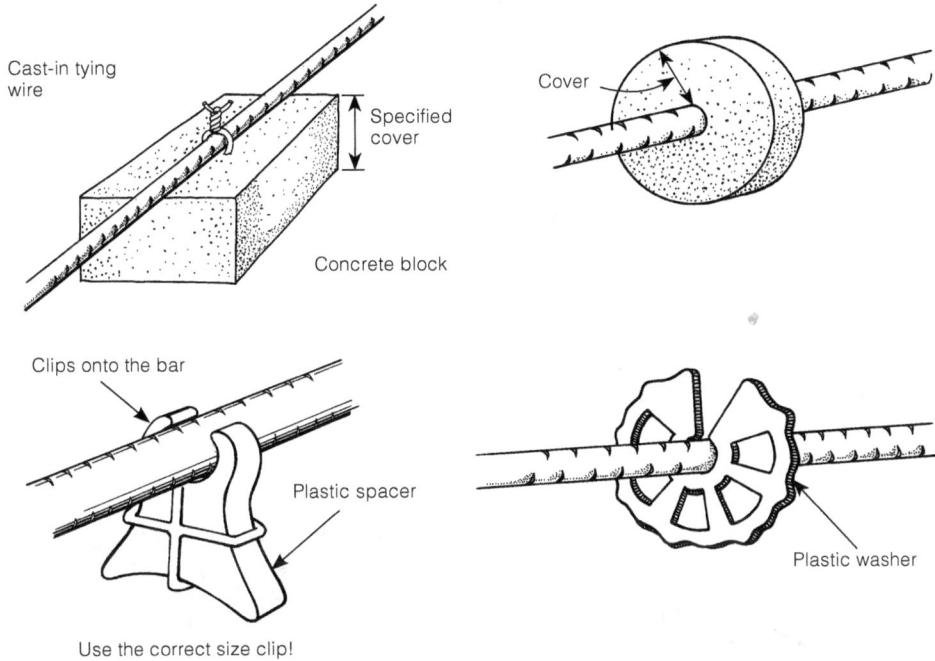

Cast-in tying wire

Specified cover

Concrete block

Cover

Clips onto the bar

Plastic spacer

Use the correct size clip!

Plastic washer

Make sure there is sufficient end cover to the bars.
Where there is reinforcement in both faces of a concrete slab, the two layers should be held at the correct distance apart by chairs tied to both layers.

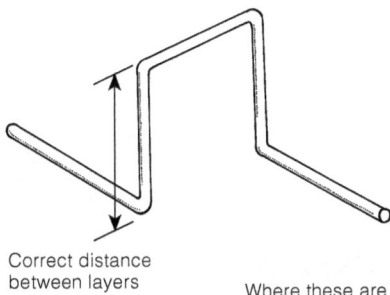

Correct distance between layers

Where these are in a floor slab they should be strong enough and close enough together to support people walking on the top steel without it sagging.

Figure 4.39

235

Reinforcement 2

Reinforcement should be accurately fixed and maintained in the position shown on the drawings before concreting is allowed to commence.

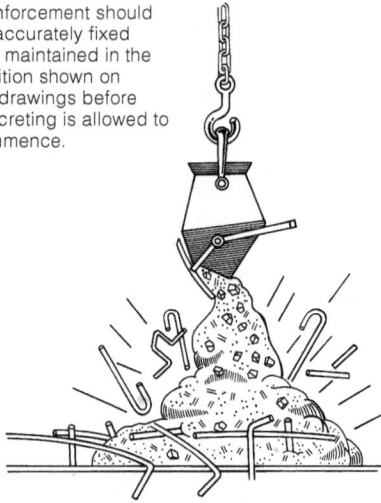

Ok

Brush off

Do not use badly pitted bars

Slight rusting is not detrimental, but brush off loose rust and scale with a wire brush as they prevent proper bonding with the concrete.

No reinforcement or tying wire should be in contact with formwork.

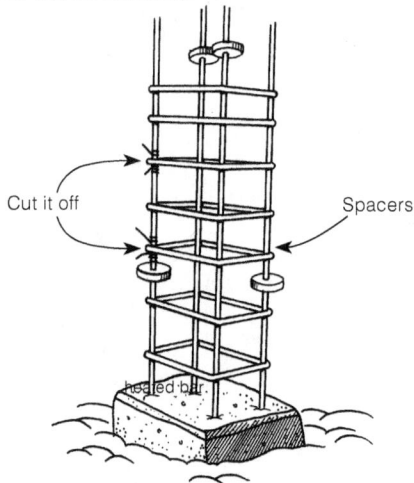

Cut it off

Spacers

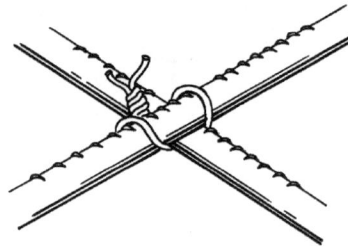

Bars should be tied together where they cross with 16 or 18 gauge soft iron tying wire at enough points to prevent movement.

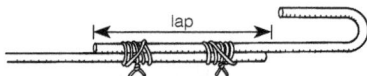

lap

Make sure that where bars are lapped that the laps are as specified on the drawings.

Before concreting, make sure that any mud or mould oil is removed.

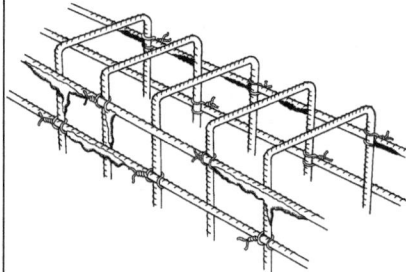

Figure 4.39

⇨

236

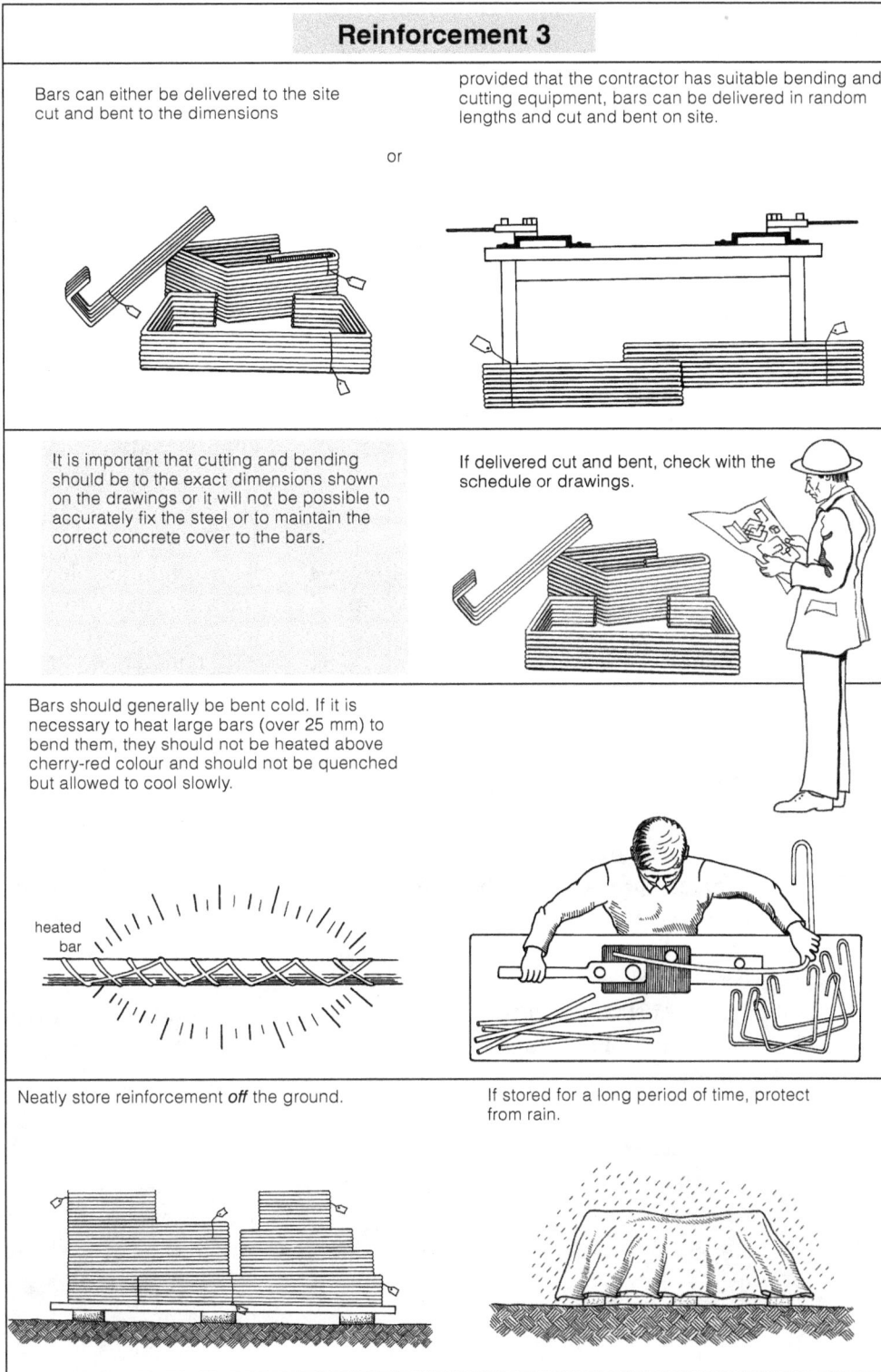

Reinforcement 3

Bars can either be delivered to the site cut and bent to the dimensions

or

provided that the contractor has suitable bending and cutting equipment, bars can be delivered in random lengths and cut and bent on site.

It is important that cutting and bending should be to the exact dimensions shown on the drawings or it will not be possible to accurately fix the steel or to maintain the correct concrete cover to the bars.

If delivered cut and bent, check with the schedule or drawings.

Bars should generally be bent cold. If it is necessary to heat large bars (over 25 mm) to bend them, they should not be heated above cherry-red colour and should not be quenched but allowed to cool slowly.

heated bar

Neatly store reinforcement *off* the ground.

If stored for a long period of time, protect from rain.

Figure 4.39

237

Activity 6

Go out on site or to a steel plant and see how many of the steel sections you can identify. Ask the site manager if you may remove some cut-off pieces to examine more closely.

4.4.8 Steel codes and specifications

These are the various codes of practice applicable to the use of steel in construction:

- SANS 10162-1: 2011 Limit state design of hot-rolled steelwork
- SANS 10162-2: 2011 Limit state design of cold-formed steelwork
- South African Institute of Steel construction handbook: 1997
- SANS 10100-1:2000 The structural use of concrete: Design
- SANS 10100-2: 1992 The structural use of concrete: Materials

4.5 Aluminium

In its natural form, aluminium is the most common metallic element found in the Earth's crust. You can find traces of aluminium in most types of rock as well as clay. However, the extraction process to obtain the aluminium is difficult and very expensive, especially when dealing with rock and clay.

4.5.1 Sources of aluminium

Bauxite is a term used to describe the aluminium ore mined for production. Bauxites can be found near igneous rocks in tropical or subtropical areas. Some bauxites occur as sedimentary or alluvial deposits. Generally, the deposits are close to the surface and can be extracted by open-cast quarrying. After quarrying, the bauxites are screened and washed to remove excess dirt. Bauxite deposits are found in Australia, southern and western Europe and the Americas. Minor deposits are found in north Africa. South Africa has to import aluminium.

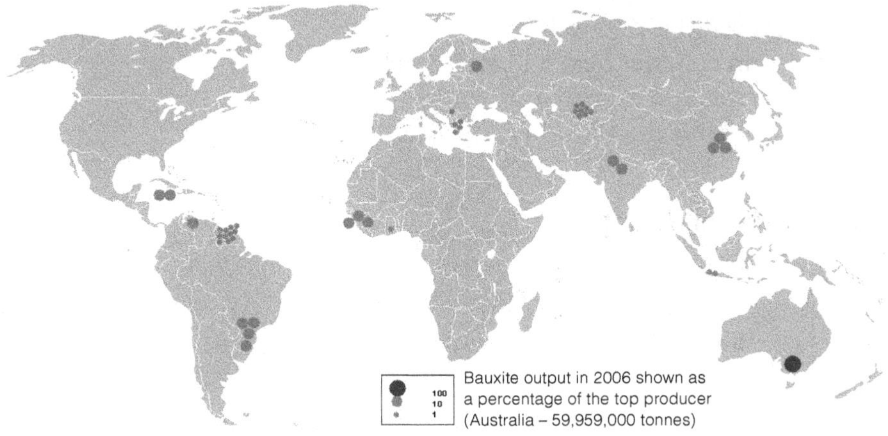

Figure 4.40 Map of the world showing the sources of bauxite (aluminium)

4.5.2 Manufacture of aluminium

The raw, washed bauxite is transported to a plant where further refining takes place. It is crushed into a fine powder and mixed with a hot solution of caustic soda. This caustic soda dissolves the bauxite into a solution of caustic soda and aluminium particles called sodium aluminate. This solution is pumped into a large tank and is stirred. Pure aluminium particles are added which attract the aluminium particles in the solution forming a clump (almost like a magnet attracts fine steel particles). This clump of aluminium particles is allowed to settle to the bottom of the tank and then the liquid solution is removed from the tank. (The liquid is stored for re-use later.) The clump of aluminium particles is passed through a rotary kiln heated to approximately 1 100 °C to evaporate any water attached to the particles. The resulting product is white powder known as **calcined alumina.** Aluminium is produced by electrolysis of purified alumina.

4.5.3 Properties of aluminium

The more important properties of aluminium are:
- It is silvery white in colour and can be very shiny (has lustre)
- Compared to other metals, it is more resistant to atmospheric corrosion

- It is very light in weight
- It can conduct heat and electricity well
- It is soft and easy to work with
- It is highly ductile and malleable.

4.5.4 Uses of aluminium

> What are the most important uses of aluminium?

Aluminium has a wide range of uses, including:
- Window frames
- Patio doors
- Corrugated aluminium roofing sheets
- Structural members
- Wires, bars, rods
- Aluminium foil
- As a fine powder for pigments, e.g. paints
- Posts, panels and balustrades
- Alloys such as copper, magnesium, silicon etc. resulting in high tensile strength and hardness while still remaining light and durable.

4.5.5 Aluminium alloy

It is very seldom that aluminium used in construction is in its natural, pure form. In most cases it is mixed with another element or elements to form a metal with a specific property, for example strength, hardness, etc. Aluminium can be mixed with a range of elements like copper, zinc, magnesium, manganese, etc.

Aluminium alloys

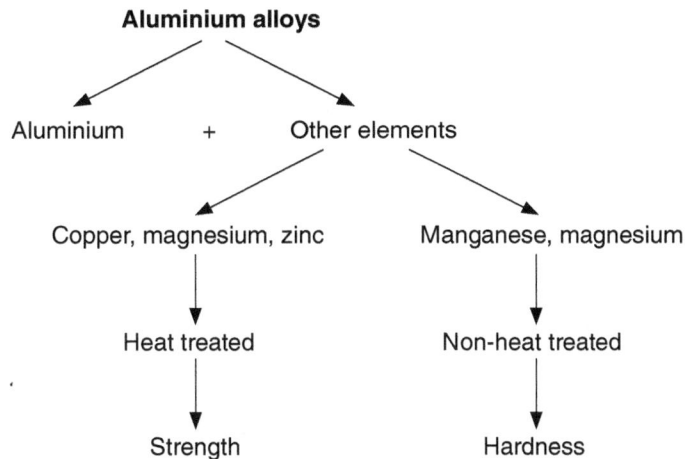

Figure 4.41 Aluminium alloys

Alloys can either be **heat-treated** or **non-heat treatable** alloys. For the former, the metal is mixed at very high temperatures and then produced in its desired shape due to the softness of the metal at the time of processing. Non-heat-treatable alloys are formed into shape by casting and do not rely on heat to complete the process. The strength and hardness of alloys are controlled during fabrication.

4.5.6 Classification of alloys

There are two accepted methods of classifying aluminium alloys: North America uses a **numeric** system, for example 617.4, and Europe uses an **alpha-numeric** system, for example A1Mg5.

Numeric system

In this system you will find a four-digit number where the first three numbers are separated by a point followed by the fourth digit, for example 617.4. Usually this four digit sequence has a preface LM denoting light metal. The first digit indicates alloy group, i.e. whether copper, silicon, etc. The second two digits identify the aluminium alloy or its purity, and the fourth digit indicates the product form.

Alpha-numeric system

In the European system, the first unit indicates the base element, for example aluminium (AL). The second unit represents the alloying group, for example Magnesium (Mg), and the third unit indicates the % of the element in the alloy, for example 5%.

Self-evaluation 4.3

1. Complete the sentences:
    a. Ferrous metals are predominantly made from
        _____.
    b. The mixture of two or more metals is called an
        _____.
    c. _____ is a term used to describe aluminium ore
        mined for production.
    d. Aluminium is more resistant to _____ corrosion
        if compared to other metals.
    e. _____ is the stress beyond which any increase in
        load causes permanent deformation.
    f. _____ is a pulling force or stress which tends to
        lengthen the member.

2. State whether the following statements are **true** or **false**:
    a. Steel is strong in compression and weak in tension.
    b. The amount of carbon present in a metal influences its
        hardness and stiffness.
    c. Mild steel can be easily welded, riveted and forged.
    d. Stress is expressed in N/mm^2.
    e. Aluminium is a good conductor of heat and electricity.
    f. Non-heat treatable alloys are heavily reliant on heat for
        fabrication.

3. Answer the following:
    a. What is steel?
    b. Define the following terms: stress, strain, compression and
        tension.
    c. What is the difference between ferrous and non-ferrous
        metals? Give examples.
    d. How would you distinguish between pieces of high carbon
        steel and mild steel?
    e. Describe the properties of aluminium.
    f. Describe the uses of aluminium.

4.6 Summary

The purpose of this unit was to:

- Describe the manufacturing, characteristics and properties of a
  number of different materials like bricks, steel, aluminium and
  timber
- Describe defects in materials
- Calculate the quantity of bricks in a wall

- Distinguish between different types of timber products, metals, steels, aluminiums and their uses
- Apply the relevant standards and codes
- Explain the relevance of environmental considerations to engineering and development.

Answers

Activity 2

Area of wall 1 $= 60 \times 3.5$m

$= 210 \text{ m}^2$

Area of wall 2 $=$ area of wall $-$ (area of windows $+$ area of door)

$= [(60 \times 3.5) - 8(2.0 \times 0.9) + (2.5 + 1.8)]$

$= 191.10 \text{m}^2$

Area of brick $= [(0.222 + 0.012) + (0.073 + 0.012)]$

$= 0.019 \ 89 \text{ m}^2$

therefore

Quantity of bricks required for wall 1 $= \left[\left(\dfrac{210}{0.019 \ 89}\right) \times 2 + 10\%\right]$

$= 23 \ 228$

($\times 2$ because it is a double wall (280 mm) plus 10% wastage)

Quantity of bricks required for wall 1 $= \left[\left(\dfrac{191.10}{0.019 \ 89}\right) \times 2 + 10\%\right]$

$= 21 \ 137$

Quantities of mortar, sand and cement:

Hint: use table 4.1 to calculate these quantities

	Wall 1	Wall 2
Quantity per 1 000 bricks	$\dfrac{23 \ 228}{1 \ 000} = 23.228$	$\dfrac{21 \ 137}{1 \ 000} = 21.137$
Cement	23.228×5 bags $= 116.10$ bags, say 117 bags	21.137×5 bags $= 105.68$ bags, say 106 bags
Sand	$23.228 \times 1 \text{ m}^3 = 23.228 \text{ m}^3$	$21.137 \times 1 \text{ m}^3 = 21.137 \text{ m}^3$
Mortar	$23.228 \times 0.6 \text{ m}^3 = 13.9 \text{ m}^3$	$21.137 \times 0.6 \text{ m}^3 = 12.68 \text{ m}^3$

Note:

- When assuming 55 bricks/single wall and 110 bricks/double and cavity wall, all the quantities will be slightly less. Try to calculate how much difference there is.
- Multiply the quantities of table 4.1 by the prices you obtained from suppliers.

Self-evaluation 4.1

1. a. Clay
   b. Frog
   c. Clamps
   d. Chuffs
   e. Course
   f. Cavity wall

2. a. False, it affects the colour by giving it a reddish tint
   b. False, it is 222 × 106 × 73 mm, the 85 mm takes into account a 12 mm mortar bedding
   c. True
   d. False, it is caused by excessive heating whereas bloating is caused by excess carbon matter
   e. True
   f. True

3. a. It is done to control shrinkage of the brick.
   b. For uniformity and ease of construction when aligning joints.
   c. To ensure that the bricks are of good quality.
   d. Yes, it is important as certain structures require bricks with high strength characteristics.

Self-evaluation 4.2

1. a. Timber
   b. Formwork
   c. Heartwood
   d. Seasoning
   e. Creosote
   f. Plywood

2. a. True
   b. False, the reverse is true, i.e. brick/outside and wood/inside
   c. True
   d. True
   e. False, softwood has needle-shaped leaves
   f. True

3. a. See fig 4.38 and section 4.3.1.
   b. There are four classifications, namely endogens, exogens, softwood and hardwood (see section 4.3.3).
   c. Use section 4.3.7 to distinguish the difference.
   d. A preservative can be applied to timber in liquid or spray form to increase the durability of the timber. The properties are defined in section 4.3.10.
   e. Plywood can be made from different types of wood and consists of layers secured (glued) together. 3 and 5 ply are normally available.
   f. They convey sap and support the tree (see section 4.3.3).

Self-evaluation 4.3

1.   a.   Iron
     b.   Alloy
     c.   Bauxites
     d.   Atmosphere
     e.   Elastic limit
     f.   Tension
2.   a.   False, steel is strong in tension and weak in compression
     b.   True
     c.   True
     d.   True
     e.   True
     f.   False, they do not rely on heat and are cast into shape
3.   a.   Steel is a mixture of various elements and is a product of pig iron.
     b.   See section 4.4.
     c.   Ferrous metals are made from iron and non-ferrous metals are not. For examples, see section 4.4.
     d.   See table 4.3.
     e.   See section 4.5.3.
     f.   See section 4.5.4.

Advanced exercises

1.   Remember the house that we are building, I now want you to calculate the number of bricks required to completely construct this house. Let's assume the following:
     a.   The house will be covered with plaster both inside and outside (including all the internal walls).
     b.   We must make allowance for all the doors (front, back, garage and internal doors).
     c.   We must make allowance for all the window openings.
     d.   We must allow for at least 10% wastage of materials used.
     e.   Remember if your house has gable ends, we must also calculate this quantity of bricks
     f.   Contact your local supplier and enquire what the cost of bricks is, then calculate the overall cost of purchase of bricks for your house.
     g.   Determine the quantities of cement and sand required to build your house using the amount of bricks you calculated

2. Using the same house construction, you need to identify where in your house you will be using timber and then calculate the following:
    a. What type of timber will be used and where as well as the quantities thereof.
    b. Contact your local supplier, obtain prices for this material and work out what the cost for the purchase of timber will amount to.
    c. Remember to keep in mind the timber wall plates, window and door frames, roof timber, built-in cupboards, etc.
    d. If you are using timber window frames, check with the suppliers in terms of the various designs available.
    e. If you are using metal or aluminium window frames, check with the suppliers what designs are available and then incorporate this into your analysis.
3. Consedering that the house will be plastered on the inside and outside, calculate what the area will be that will require plastering. Once this is determined, identify what quantities of cement and sand will be required should you use a 1:6 mix.

5

Environmental issues

Learning outcomes

After studying this unit, you should be able to:

- Explain the relevance of environmental considerations to engineering and development
- Discuss the interdependence of conservation and development
- Raise environmental consciousness
- Identify the causes of soil erosion, water pollution, toxic waste, air pollution and know how to prevent them
- Describe sewage treatment and eutrophication
- Understand the interaction between civil engineering and the environment.

5.1 Environmental engineering

South Africa is a rich, but still developing country. We have an abundance of resources, yet many people have a low standard of living, and population growth is rapid. There is great pressure to increase the pace of development. Environmental engineering is a specialist discipline that complements civil engineering. This unit does not discuss any topics in great detail, but provides the essential background information for this course.

There are many ways to interpret what is meant by the term environment. In many cases it is an all-encompassing term used to describe or understand our surroundings, their effects and the impact of our interactions. A general description as applied by ISO is 'the surroundings in which an individual or organization operates, including air, water, land, natural resources, flora, fauna, humans and their interrelation' (ISO14001: 2004).

What is the environment?

5.2 The interdependence of conservation and development

Generally, there should be no conflict between **conservation** and **development**. It is speculated that better conservation efforts will help South Africa to meet its developmental goals and, at the same time, efficient development will ensure that conservation goals will also be met. For example, soil conservation is an important part of agricultural development: the more soil that is lost, the less food can be grown. By the same token, if there is little agricultural development, then more soil will be lost because of poor farming methods.

An economy cannot thrive if environmental resources are lost or badly managed. The health of the South African economy is largely dependent on the renewable resources important to such industries as forestry, fishing, agriculture and tourism. Wildlife conservation, for example, stimulates the tourist industry, and marine conservation prevents the collapse of the fishing industry.

How can economic development help conservation?

Environmental resources will also be lost or badly damaged if the economy does not thrive. For example, unemployed or impoverished people will use resources around them, such as forests and wild animals, to survive; industries in poor countries may ignore pollution dangers to produce cheap goods.

Figure 5.1 Industrial pollution

Economic development helps conservation, for example development of the tourism industry provides both income and incentives to protect wildlife resources. Efficient development programmes generate money and can support conservation programmes. Conservation efforts cannot be effective if money is not available, or if the economy is not able to satisfy the basic needs and aspirations of the majority of the population. Where there is a great want, there is always going to be over-exploitation of resources.

There are three major causes of over-exploitation: poverty, ignorance and greed. To some extent, ignorance and greed can be handled through education and regulations, but only economic development can alleviate poverty.

To be successful, economic development must ensure benefits for the current and succeeding generations through a careful balance of development and conservation.

5.3 Population figures

Extraordinary population changes have taken place in the past 150 years. If present trends continue, there will be at least 8.5 billion people by the year 2025.

Clearly, more people make greater demands on the Earth's resources. However, human impact on the Earth is not only determined by numbers of people, but also by how much energy and other resources each person uses. Sustainable living is possible only if human numbers and demand for resources are kept within the Earth's carrying capacity.

Giving people the means (through the vote, improved social and legal status, education, access to family planning and financial independence) to choose the size of their families will not only help keep the population in balance with resources, it is also a way of assuring, especially for women, the basic right of self-determination.

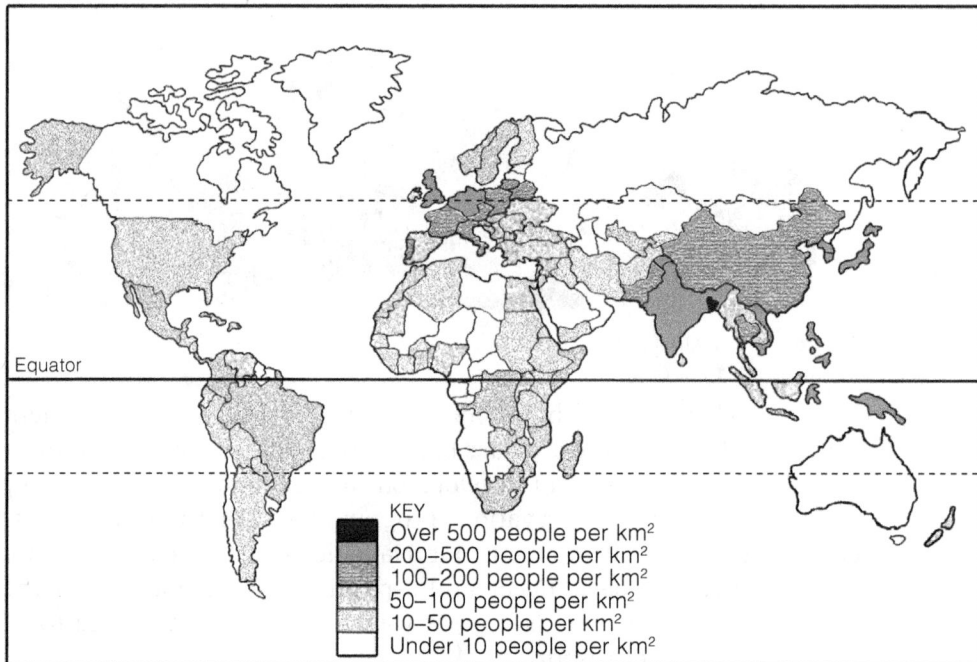

Figure 5.2 World population density

There are a number of environmental problems, for example soil erosion, water pollution and air pollution. Lets look at a few of the concerns and how we can deal with them.

5.4 Soil erosion

Soil erosion is a natural process. It only becomes a problem when human activity causes it to occur much faster than under natural conditions. The annual soil loss in South Africa is estimated at 300–400 million tonnes, nearly three tonnes for each hectare of land. This is equivalent to about 150 000 hectares each 150 mm deep (1 hectare equals 10 000 m^2). Put another way, for every tonne of maize, wheat, sugar or other agricultural crop produced, South Africa loses an average of 20 tonnes of soil.

5.4.1 Causes of soil erosion

How does soil erosion happen?

Wind and water are the main agents of soil erosion. The amount of soil they can carry away is influenced by two related factors:

- **Speed:** The faster either moves, the more soil can erode
- **Plant cover:** Plants protect the soil and in their absence wind and water can do much more damage.

Erosion removes the topsoil first. Once this nutrient-rich layer of soil is gone, few plants will grow. Without soil and plants, the land becomes desert-like and unable to support life – a process called **desertification**. It is very difficult to restore desertified land.

Other ways soil becomes unusable are **salination,** where minerals like sodium chloride (salt) are washed down from industries and sewage into the soil causing it to become alkaline, and by the invasion of alien vegetation.

Figure 5.3 The activities leading to land pollution and degradation

251

5.4.2 Preventing soil erosion

The prevention of soil erosion requires political, economic and technical changes. These include:

- Incentives to encourage farmers to manage their land sustainably
- The use of contour ploughing and wind breaks. However, contour ploughing does not eliminate soil erosion completely
- Allowing indigenous plants to grow along river banks instead of ploughing and planting crops right up to the water's edge
- Encouraging biological diversity by planting several different types of plants together
- Conservation of wetlands.

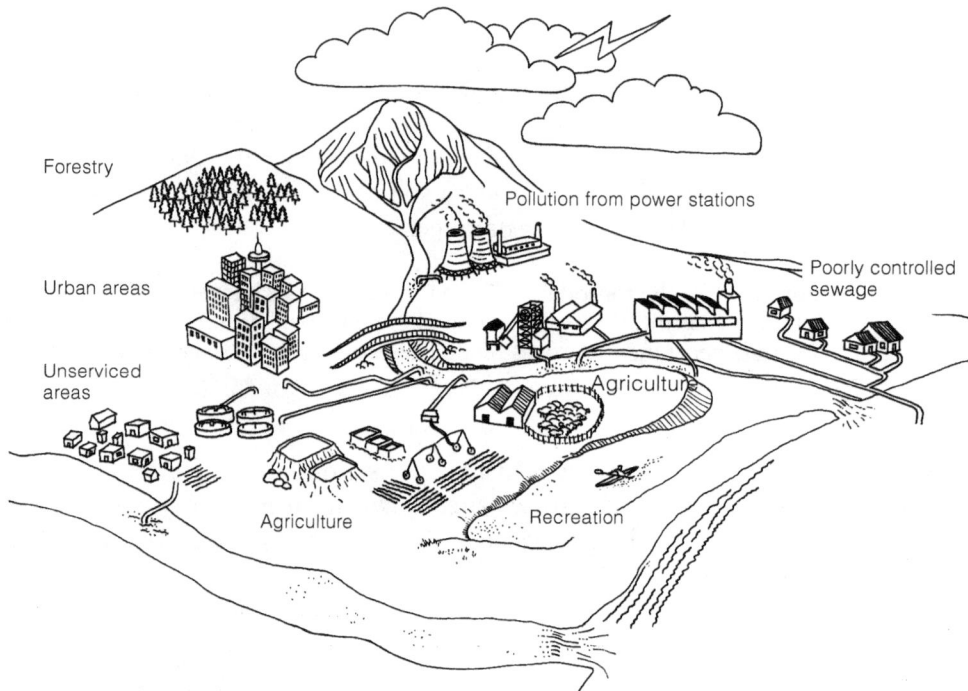

Figure 5.4 Activities leading to water pollution and consumption

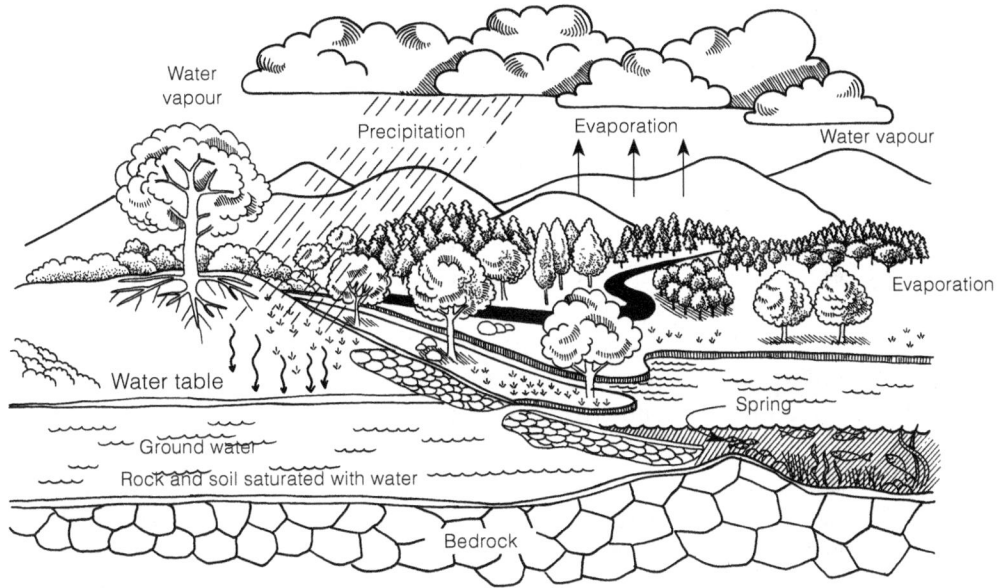

Figure 5.5 The water cycle

Activity 1

In a group, spend about 20 minutes discussing:
- The various forms of soil and land pollution
- The causes, i.e. natural or artificial
- The effects
- The remedies.

Use a large blank sheet of paper and draw a flowchart to help explain your reasoning. At the end of the discussion see which group has the best chart and make a copy to use for future reference.

5.5 Water pollution

The way land is used has a marked effect on water. Changing the use of land can increase the potential for flooding or reduce the amount of water flowing into rivers. It is not only the **quantity** of water that can be changed but also the **quality**. When rainwater falls on land, the quality of the water may be changed in many ways. In fact, even as rain falls through the air, it can be contaminated by air pollution.

If, for example, an area is developed as a forestry plantation, especially the higher **catchment areas**, people living downstream will notice a reduced flow of water in the river. If a large tract of land is over-grazed, the risk of flooding will increase and the reservoirs will be silted up faster because there

253

is no vegetation to retard soil erosion. A **catchment** is the entire area that is considered when calculating drainage. So in this case, higher catchment areas are those located at a higher level, for example hills or mountains.

When large quantities of water are used for irrigation, apart from reducing the flow in the rivers, the water which finds its way back to the river often carries with it large amounts of fertilisers and chemicals which will affect the quality of the river water. Cities, residential communities, industrial areas and mining all have an impact on water quality and quantity.

As all water eventually flows into the sea, contaminated water pollutes water in estuaries and near beaches. This can affect tourism, recreation and the breeding areas of fish.

The present water Act does not regard the environment as a legitimate user of water. Recently the needs of the environment have been increasingly taken into account and the impact of development is beginning to be considered seriously.

The environment is now being regarded as a user of water, competing with other users such as industry, agriculture and municipalities. The law needs to view the environment as the resource base from which all development leads – the foundation on which all else depends.

Do you know who the current Minister of Water Affairs is?

The existing Act provides the Minister of Water Affairs with far reaching powers to restrict water usage during times of water shortage and drought. It does not, however, provide measures to ensure the conservation of water at all times. Although water conservation can be induced through tariff policies in some cases, many persons who enjoy water rights are not subject to any tariffs or levies on the water they use, for example farmers and residents of informal settlements. Powers to enforce water conservation measures are therefore needed, particularly as South Africa is situated in a semi-arid region.

Have you noticed rivers or dams covered with plant growth? Large dams in Gauteng, for example Vaal Dam and Hartebeesport Dam, are covered with a broad-leaved water plant called water hyacinth. This plant affects the quality of the water by extracting oxygen, thereby suffocating other living organisms. Clean-up campaigns have been started to clear water hyacinth from South Africa's waterways.

5.5.1 Causes of water pollution

There are many causes of water pollution including:
- Floating matter – sawdust, refuse, plastic objects
- Matter in solution – chemicals, acids
- Toxic matter – insecticides, chlorinated hydrocarbons
- Micro organisms – bacteria, viruses
- Radioactive material – nuclear power stations
- Heat from industries, which affects fish that can only live at certain temperatures
- Unpleasant odours caused by the decomposition of industrial effluent
- Matter causing discolouration – dyes, detergents.

Did you know that three billion tonnes of soil are displaced from cultivated land each year?

Sedimentation

As sedimentation absorbs light energy, silt particles may increase the water temperature. When silt sinks to the bottom of a stream it can smother plant life. Remember that photosynthesis depends on the absorption of sunlight and if this is interrupted in any way, reproduction cannot take place.

Acid rain

The natural cycle of evaporation and rain is the same as a laboratory process of purifying water. But, unlike controlled lab processes, rain droplets absorb impurities in the atmosphere. Natural gases dissolve in this moisture to form acids. Sulphur compounds emitted from volcanoes and certain industries interact with moisture, to form sulphuric acid (a high pH value), resulting in what is commonly known as **acid rain**.

Another form of acid rain is the mixture of carbon dioxide and water droplets in the atmosphere. The water and the carbon dioxide react to form carbonic acid which then falls to the ground as acid rain. Soil does, however, contain naturally occurring substances, such as magnesium and calcium, that act as neutralisers of these acids.

Acid rain causes an increase in the rate at which calcium and other nutrients in the soil are used up. It also leads to the death of fish due to the higher acidity of the water and a reduction in forest growth as a result of lack of nutrients in the soil.

Activity 2

In a group, spend 20 minutes discussing:
- The various forms of water pollution
- The causes, i.e. natural or artificial besides those mentioned above
- The effects of these causes
- What can be done to remedy the situation.

Use a large blank sheet of paper and draw a flowchart to help explain your reasoning. At the end of the discussion see which group has the best chart and make a copy to use for future reference

5.6 Toxic waste

Figure 5.6 A nuclear power station (left) and coal-fired electricity towers (right)

Toxic waste is a widely used term that is difficult to define. It includes substances that are harmful to life and the environment, i.e. waste with any of the following characteristics:
- Chemically and biologically poisonous
- Radio-active
- Flammable
- Explosive
- Corrosive
- Carcinogenic (causes cancer)
- Mutagenic (damaging chromosomes)

- Teratogenic (causing defects in the unborn)
- Bio-accumulative (accumulating in the food chain).

Toxic or poisonous wastes are produced during industrial, chemical, medical and agricultural processes. Even household, office and commercial wastes contain small quantities of toxins (e.g. old batteries, pesticides and their containers). **Dioxins** are produced in the burning of chlorine containing substances, for example plastics, the manufacture of iron and steel, and some organic chemicals like herbicides. They are also found in leaded petrol and bleached white paper. **Heavy metals** have widespread industrial use, including gold extraction (arsenic) and are found in batteries (mercury, cadmium, lead) and leaded petrol. **Radio-active waste** is created by nuclear power generation and certain medical procedures, for example X-rays.

5.6.1 What can be done about toxic waste?

The ultimate solution to toxic waste is to reduce its production by:
- Substituting non-polluting alternatives, e.g. replacing the chlorine used to bleach wood with oxygen
- Creating efficient production processes and maintaining machinery to reduce waste production
- Recycling waste, thereby reducing pollution and saving costs, e.g. expensive, toxic heavy metals could be re-used
- Not using or producing what cannot be recycled or is too dangerous to handle in the first place.

As an individual you should:
- Report dump sites to your nearest health officials, your local environment organisation, the Department of Water Affairs or the Institute of Waste Management
- Avoid using toxic products at home
- Dispose of household toxic substances into plastic bags in municipal bins, not down the toilet or into the drains at home, stormwater drains, rivers or dams
- Become informed, join forces with concerned people and make your voice heard.

Activity 3

In a group, spend about 20 minutes discussing:
- The various forms and products of recycling
- Ways and means of reducing the amount of waste
- The benefits of recycling
- How your group would go about creating educational awareness within your community with regards to recycling.

Use a large blank sheet of paper and draw a flowchart to help explain your reasoning. At the end of the discussion see which group has the best chart and make a copy to use for future reference. Why not start a competition by creating posters and 1st prize can go to the best poster, with adjudication being done by the whole class?

Activity 4

Get hold of a bag of household waste, either from home or from a waste bin nearby. Identify all the items in it that can be recycled and separate them out. Items like plastics, paper, cans and glass can all be recycled. In a group, discuss how much is wasted everywhere – even at home. Think about all the waste from the households in your area, a shopping complex or industrial area. Where does all this get disposed? How much of this waste can be reduced by recycling? Think about the financial benefits of recycling as well as the possibilities to create employment.

Do a presentation to the class highlighting what your group discussed.

5.7 Air pollution

Global warming describes the gradual increase of the air temperature in the Earth's lower atmosphere. The term **greenhouse effect** describes the warming effect that certain gases have on the temperature of the Earth's atmosphere under normal conditions.

Human population growth and industrial expansion have led to greater air pollution and a change in the composition of the Earth's atmosphere. Some pollutants enhance the natural greenhouse effect, resulting in increased global atmospheric temperatures.

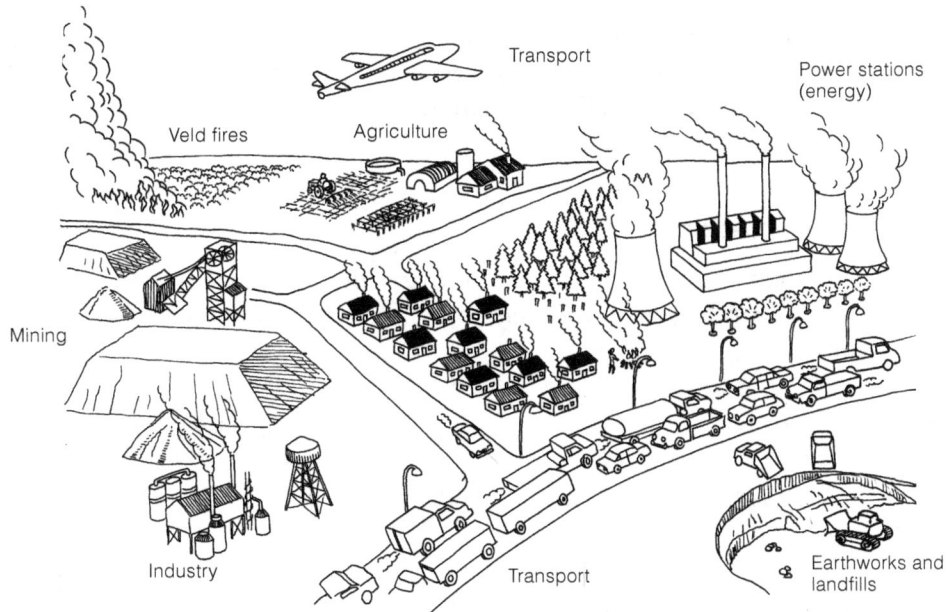

Figure 5.7 Activities leading to air pollution

The main greenhouse gases are:

- Water vapour
- Carbon dioxide (CO_2), the pollutant most responsible for increased global warming. It is released into the atmosphere mainly through the burning of fossil fuel (e.g. coal, petrol and diesel). The widespread destruction of natural vegetation, particularly forests, has contributed to increased atmospheric CO_2 levels.

Can you name the main green house gases?

- Methane (CH_4), which doubled in concentration between 1750 and 1990, mainly as a result of agricultural activities
- Nitrous oxide (N_2O), also a product of burning fossil fuel, has increased by 8% over the same period
- Chloro-fluorocarbons (CFCs), which damage the ozone layer and are

259

potent greenhouse gases. Their concentrations in the atmosphere increase by about 4% every year. Some older fridges and aerosol spray cans contain CFCs which, when released into the atmosphere, cause damage.

5.7.1 Ozone depletion

The Earth's atmosphere has many layers, but the one most important to us is the stratosphere, which is approximately 40 km above the Earth's surface.

Do you still remember how the earth's atmosphere is made up and how many layers there are?

By now you must have heard of CFCs as there has been quite an outcry in recent years regarding the emission of CFCs into the atmosphere. CFCs used to found in spray cans, like those containing deodorant and paint, and in some fridges. CFCs are an air pollutant that reacts with oxygen in the stratosphere and, together with other groups of elements, continuously generates and reacts with oxygen elements, thereby reducing the amount of oxygen molecules in the stratosphere. This reduction causes the stratosphere to become thinner, resulting in the 'hole' in the atmosphere to which we commonly refer. This, in turn, causes an increased incidence of skin cancer and accelerates production of greenhouse gases. Scientists have measured this thinning and found the stratosphere to be thinnest over the North and South Poles. This will result in:

- Less protection against harmful UV rays
- Global warming
- Oxygen depletion in the atmosphere.

Air pollution can be reduced by encouraging:

- A reduction in the consumption of electricity
- The use of lift clubs, public transport, bicycles and walking
- Reducing, reusing, recycling to save energy – the manufacture of all products requires energy.

Activity 5

In a group, spend 20 minutes discussing:
- The various forms of air pollution
- What/who are the causes
- The effects of each on the environment
- What control measures are used
- What can you do?

Do a presentation to the class highlighting what your group discussed.

5.8 The balance between humans and nature

There are various ways in which the actions of humans damage the inter-relationship between humans and nature.

Due to our poor knowledge of the interacting web of habitat and life, the impact of agriculture, road building, construction and open-cast mining operations on large stable ecosystems is not always anticipated. Such activities can deprive communities of environmental amenities and services, which leads to a reduction in the quality of life, not only for humans but also of plants and animals.

An example of the destruction of the environment is the construction of a new housing layout. The site is first cleared and levelled. This involves the removal of all vegetation and topsoil, which results in the loss of species of plants and the wiping out of the insects and birds that fed on them. This kind of destruction can be reduced by an environmental impact study carried out by experts in their field, for example botanists who can identify threatened species and take the necessary steps to protect them.

5.9 Sewage

Definitions

Sewers are underground pipes and channels that convey **sewage**, i.e. waterborne community waste like toilet, bath, kitchen and industrial waste from factories.

Sewage is treated to make it:
- Non-injurious to health
- Inoffensive to sight and smell
- Safe to discharge into the water systems so that it does not harm ecosystems naturally present
- Fit for re-use.

5.9.1 Sewage treatment

The main sewage treatments are:

- **Biological:** Living organisms break down the sewage, i.e. micro-organisms (bacteria, protozoa, algae). The oxygen necessary for this process is obtained directly from the atmosphere or fed through pipes in the tanks where the sewage is held, called an activated sludge plant
- **Mechanical:** Pumps, screens, aerators help in the removal of solid waste in the way of rags, plastic, sand, stone, etc. Aerators mix the sewage introducing oxygen mechanically
- **Chemical:** Oxidation is improved chemically by mixing certain chemicals which, when dissolved, increase the oxygen content, allowing for further biological breakdown. Injection of chlorine into the clean, treated water kills harmful bacteria.

5.9.2 Phases in sewage treatment

Primary treatment

This is the most elementary process where sewage is physically processed. Sewage entering the sewage works through a very large diameter pipe (approximately 1 300 mm diameter) will be:

- **Screened,** where the sewage passes through a screen to remove coarse pieces of solid matter that is lighter than grit, e.g. paper, plastic, rags, etc. It will then move to:
- The **degritters,** to remove inorganic materials such as stones, metal, glass, etc. from where it will then move on to:
- The **primary sedimentation tanks,** where settlement of impurities in the form of liquid sludge occurs, which will then be directed to the area where the liquid sludge will be thickened and eventually settle to the bottom of the tank. The settlement of impurities is aided by the addition of certain chemicals called flocculants which cause coagulation of impurities and thereafter settlement. The thickened sludge is directed to:
- **Digestors,** dome-shaped enclosed tanks from which air is excluded, where anaerobic organisms attack the raw sludge and break it down further.

Secondary treatment

Water is transported from the primary sedimentation tanks to the **aeration tanks** where oxidation takes place. This causes micro-organisms to break down the polluting matter further. From here it is transferred to the **secondary sedimentation tank** where further settlement takes place. The sludge which emerges from here is transferred to the **filtration plant** where percolating filter presses dewater it (remove water or liquid from

262

solids) under pressure between special filter cloths. A 'cake' remains on the cloth and is removed to be incinerated or buried.

Tertiary treatment

Water passes from the secondary digestors to the **maturation ponds** (a series of ponds, usually five) which cover a large area. Here the water will pass from each in sequence, after settlement has taken place. Before the water is passed into the ground, a river or the sea, it will be purified by chlorination through a chlorination plant.

The **methane gas** produced by primary digestors is stored in a gas holder. This flammable gas can be used in various situations, for example heating water for the digestor or incinerating dried sludge which is removed from the **sludge drying beds**.

5.10 Eutrophication

Eutrophication is the increase by natural means of nutrients such as nitrogen, phosphorous and carbon in watercourses (dams, rivers) thereby increasing the growth of algae with the resultant decrease in water quality.

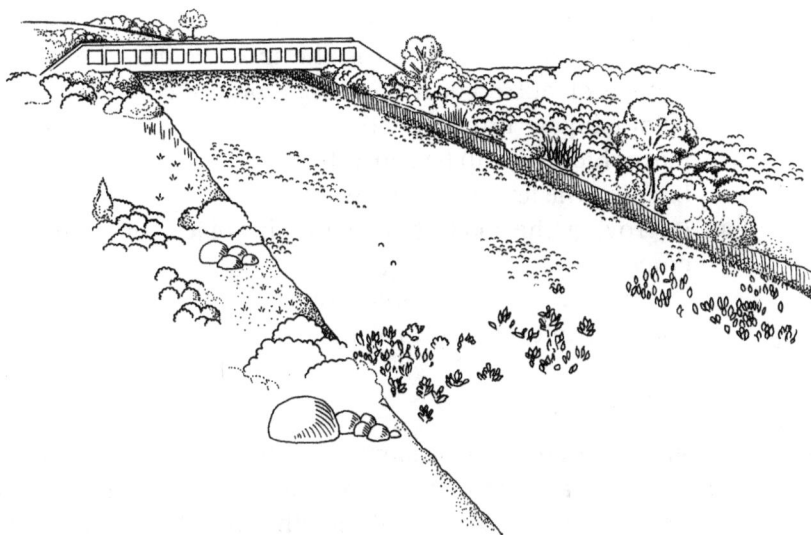

Figure 5.8 A dam or river course overgrown with water hyacinth (once fully established, they cause the water quality to deteriorate and fish die due to oxygen starvation. The Vaal Dam is a good example of the problems experienced with plant growth in the watercourses)

The natural source of nutrients in watercourses is through weathering and the leaching of the substrates into the river. Thus vegetation, geology features and rain play a role in the natural process. The artificial source is as a result of agricultural activities, which alter the pathways and rate of nutrient transport.

Have you noticed that some rivers and dams are overgrown with water hyacinth, which flourishes where there is an abundance of nutrients?

As a result of increased nutrients in the water, biological activity increases. This is a stimulant for the whole food chain, for example plants, which in turn affects animals.

The adverse effects of this increase in nutrients include increased purification costs, interference with the recreational use of lakes, water taste and odours, toxins produced by the algae and a possible decrease in land value.

Eutrophication increases the biomass of plants and animals in a water body. Thus, fish exploitation is the most common method of taking advantage of eutrophication. Irrigation water is also improved as a result of a higher nutrient content.

Eutrophication can be controlled by:

- Limiting the fertility of the water
- Improving the food chain with the object of producing valuable crops
- Controlling unwanted organisms
- Using chemicals to control algae.

Activity 6

Divide the number of students in your class by half. Use the lecturer as the mediator and enter into a debate where the one half of the class is for conservation and the other half for development. Discuss issues around these topics:

- Pros and cons of conservation and development
- Why should developers consider conservation?
- Technology versus natural resources
- Conservation and development in third world countries compared with first world countries
- What is happening in South Africa with respect to development and conservation?

Self-evaluation 5.1
1. Complete the sentences:
    a. _____ relate to anything that can be used to improve or maintain well being.
    b. There are three causes of over-exploitation, namely _____, _____ and greed.
    c. _____ _____ consists of substances that are harmful to life and the environment.
    d. Global warming describes the gradual increase of _____ _____ in the Earth's atmosphere.
    e. _____ are underground pipes and channels that convey waterborne community waste.
    f. _____ is the increase of nutrients in water courses promoting the growth of water plants.
2. State whether the following statements are **true** or **false**:
    a. Conservation is the condition in which basic needs and aspirations are met.
    b. Soil erosion can be slowed down by means of planting shrubs and trees.
    c. Water that falls from the sky as rain is unpolluted.
    d. Medical waste can also be classified as toxic waste.
    e. The rising of the water level of the oceans can be attributed to global warming.
    f. Eutrophication increases the water quality and reduces plant growth.

How does environmental engineering affect civil engineering and construction?

Because civil engineering and construction interact very closely with the environment, it has now become customary that any civil engineering project has to include an environmental impact study (EIS). This entails an investigation to assess the impact of the proposed project on the environment, especially in environmentally sensitive areas such as wetlands, nature reserves and protected areas.

When doing an EIS, specialists in their respective fields are consulted and their findings are reflected in an environmental impact report.

Usually this document is made available to all stakeholders and often a public meeting is arranged to allow members of the public to express their views.

The environmental impact report may then form the basis for decisions that are taken at the design stage of the civil engineering project. It is important that the environment is protected at all times and the impact of the proposed project is minimised. It is therefore important that the engineer takes cognisance of environmentalists.

The role of civil engineering in promoting environmental engineering

You may have realised that engineering and development go hand-in-hand. From an engineering perspective, attention during the design and construction phases must be given to:

■ Minimising pollution (water, air, soil, etc.)
■ Reducing waste and possible recycling
■ Minimising risk to the environment
■ Soil erosion
■ Depletion of resources.

All the above and more must be considered in order to achieve sustainable development

Activity 7
Can you think of activities during the construction phase of civil engineering projects that could result in potentially harmful or beneficial impacts on the environment? Use this discovery as an individual project or discuss in class.

5.11 Summary

This purpose of this unit was to:
■ Discuss the interdependence of conservation and development
■ Raise environmental consciousness
■ Identify the causes of soil erosion, water pollution, toxic waste and air pollution and know how to prevent them
■ Describe sewage treatment and eutrophication
■ Understand the interaction between civil engineering and the environment.

Answers

Self-evaluation 5.1
1. a. Resources
   b. Poverty, ignorance and greed
   c. Toxic waste
   d. Air temperatures
   e. Sewers
   f. Eutrophication
2. a. False, conservation is the preservation of natural resources for future generations
   b. True
   c. False, this is often not the case, e.g. acid rain
   d. True
   e. True
   f. False, eutrophication decreases the quality of water and increases plant growth

Advanced exercises

1. Review the entire cement and bitumen processes, i.e. from raw materials to finished product, and compare how they each impact on the environment. Also look at the legal implications relating to the environmental impacts. Aspects that you must consider are:
   a. Recovering raw materials and transport thereof
   b. The manufacturing process and packaging – again look at transport
   c. Use of the manufactured product by the user
   d. Sustainability and re-use
2. Define and explain the term 'bio-sphere'
3. Define and explain the term 'biodiversity'
4. How do the biosphere and biodiversity relate to the environment?

Laboratory practicals

Throughout this handbook are tests that you may be exposed to during this semester of study. Some of these tests are fairly simple to perform whilst others are more complex in nature. To assist you in doing the tests the correct way, some of the more important tests procedures are described in the units. However, to gain more valuable insight into laboratory procedures and tests, it is best to consult the handbooks mentioned in the text, i.e. TMH1, ASTM, etc.

There are a number of reasons why it is necessary to do tests, but the more important ones are:

- To provide a record of control
- To ensure that standards are maintained
- To ensure correctness of the product
- To ensure quality control.

Unfortunately, it is impossible to describe all the tests necessary, but the important ones have been simplified to assist in your understanding and enhance your learning.

6

Internal building construction materials used for finishes

Learning outcomes

After studying this unit, you should be able to:

- Understand other engineering materials used as finishes
- Identify types of floor finishes
- Identify types of wall finishes
- Identify types of paint and finishes.

We will not try to identify all the finishing materials available as the list can be rather extensive, but the intention is to focus on the basics and any additional materials or features can be researched separately.

6.1 Introduction

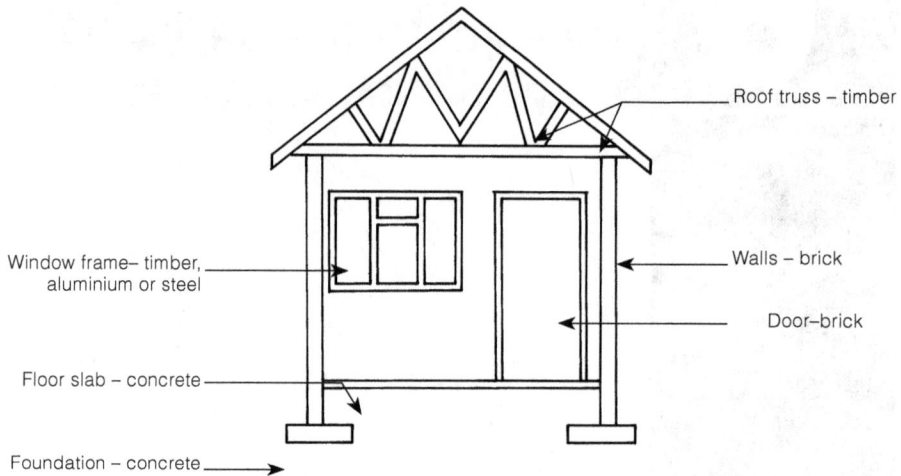

The house needs to be liveable and in order to achieve this, you need to go through each and every room to identify its purpose and thereby determine what finishes are required. Examples of this type of decision making are as follows:

- In the kitchen, the walls need to be painted, kitchen cupboards attached, tiles placed on the floors and against the walls, etc.
- In the bedrooms, there are choices of floor covering, e.g. carpets, tiles, laminated floors, vinyl floors, whilst painting also has to be done.

So we can go through each of the rooms and identify what finishes we require.

6.2 Floor coverings

There are different types of floor coverings that can be used, ranging from:

- Textile (carpets or carpet tiles)
- Tile flooring (clay, stone, metal, quartz, etc.)
- Timber floors (interlocking pine flooring, laminated floors, etc.)
- Vinyl floor coverings.

6.2.1 Ceramic tiles

Ceramic tiles are manufactured from clay materials that are quarried and formed in a mould. Usually a **dry press method** involves a mixture of dry material being pressed into a mould under extreme pressure. An **extruded method** is similar, except that the material is slightly wet when compressed into the mould. A **slush mould** is when the material is very wet when poured into the mould and hardened by means of baking in a kiln at extremely high temperatures.

6.2.2 Porcelain and non-porcelain tiles

A ceramic tile can either be classed as a porcelain or a non-porcelain tile. Porcelain tiles are made from clay and other minerals as well as a white dust (sand) called feldspar. Feldspar is a type of crystal found in rock that acts as a flux during the kiln drying process, melting into a glass-like material which bonds all the ingredients together. Modifications to the ingredients and the kiln drying process (e.g. temperature) can create enormous variety in the appearance and characteristics of the final product.

Porcelain and non-porcelain tiles can either be **glazed** or **unglazed**. A glazed tile has a matte, semi-gloss or high gloss finish applied to the surface during the manufacturing process. Usually the tile is placed in the kiln (oven) a second time to harden the glaze after it has been applied. Glazed tiles have increased stain resistance, scratch resistance and traction as well as decreased water absorption when compared to an unglazed tile.

Non-porcelain, ceramic tiles are among the most economical types of flooring. Porcelain tiles are normally more expensive and harder to work with than non-porcelain tiles.

6.2.3 Natural stone tiles

Nature stone tiles are produced from natural material that is quarried, slabbed, finished and cut to size. Common types of stone used for natural stone floor tiles are granite, marble, limestone and slate – each presenting thousands of varieties and characteristics, depending on where and when the stone was quarried.

6.2.4 Granite

Granite is a type of igneous rock that is very dense and hard. It has distinctive speckled minerals within the rock which influence its appearance. Granite is nearly impervious and once polished, resists scratching. Generally these tiles are used in kitchens and in high-traffic areas, but granite is also used for its aesthetic appeal.

6.2.5 Marble

Marble is a type of metamorphic rock that is rich in veining and is available in a variety of colours. Marble however is more porous than granite and is not recommended for kitchen flooring unless polished and sealed on a regular basis.

6.2.6 Limestone

Limestone is a type of sedimentary rock that offers an earthy appearance in both light and dark shades. Its surface can be textured or polished smooth but unlike granite, limestone is less dense and can stain and scratch easily. It is not recommended for kitchen or high-traffic flooring applications.

6.2.7 Travertine

Travertine is a type of limestone that offers an unusual crystallised appearance with an earthy tone. Travertine is very soft and porous with a natural surface that is pitted – its surface can however be polished to improve its appearance. As with limestone, travertine tiles can scratch and stain easily and often special care is required to maintain their appearance.

6.2.8 Slate

Slate tiles are a form of metamorphic rock that is extremely dense and durable. They are available in darker, earthy tones and their surface is naturally textured but can be smoothed and polished. Due to its hardness and durability, slate is recommended for kitchens and highly trafficked areas.

Figure 6.1 Various types of tiles used in construction
Source: http://www.natural stonewarehouse.com

6.3 Painting

In the olden days, bushmen paintings depicted artwork done in caves which is still visible today. These paintings were done using a red or yellow ochre, manganese oxide or charcoal.

Today and with technological advancement in paint manufacture and techniques, we have a wide variety of different paint types.

6.3.1 Binder, vehicle or resins

The binder, commonly called the vehicle, is the film-forming component of paint. It is the only component that must be present. Components listed below are included optionally, depending on the desired properties of the cured film.

The binder imparts adhesion and strongly influences such properties as gloss, durability, flexibility and toughness.

Binders include synthetic or natural resins such as alkyds, acrylics, vinyl-acrylics, vinyl acetate/ethylene (VAE), polyurethane, polyesters, melamine resins, epoxy or oils. Binders can be categorised according to the mechanisms for drying or curing. Although drying may refer to evaporation of the solvent or thinner, it usually refers to oxidative cross-linking of the binders and is indistinguishable from curing. Some paints form by solvent evaporation only, but most rely on cross-linking processes.

Paints that dry by solvent evaporation and contain a solid binder dissolved in a solvent are known as lacquers. A solid film forms when the solvent evaporates, and because the film can re-dissolve in solvent, lacquers are unsuitable for applications where chemical resistance is important. Performance varies by formulation, but lacquers generally tend to have better UV resistance and lower corrosion resistance than comparable systems that cure by polymerisation or coalescence.

Latex paint is a water-borne dispersion of sub-micrometre polymer particles. The term 'latex' in the context of paint simply means an aqueous dispersion; latex rubber (the sap of the rubber tree that has historically been called latex) is not an ingredient. Latex paints cure by a process called coalescence where first the water, and then the trace, or coalescing, solvent, evaporate and draw together, softening the latex binder particles and fusing them together into irreversibly bound networked structures, so that the paint will not redissolve in the solvent/water that originally carried it.

Recent environmental requirements restrict the use of volatile organic compounds (VOCs), and alternative means of curing have been developed, particularly for industrial purposes. In UV curing paints, the solvent is evaporated first, and hardening is then initiated by ultraviolet light. In powder coatings, there is little or no solvent, and flow and cure are produced by heating of the substrate after electrostatic application of the dry powder.

6.3.2 Diluent or solvent

The main purposes of the **diluent** are to dissolve the polymer and adjust the viscosity of the paint. It is volatile and does not become part of the paint film. It also controls flow and application properties, and in some cases can affect the stability of the paint while in liquid state. Its main function is as the carrier for the non-volatile components. To spread heavier oils (e.g. linseed) as in oil-based interior housepaint, a thinner oil is required. These volatile substances impart their properties temporarily – once the solvent has evaporated, the remaining paint is fixed to the surface. Water is the main diluent for water-borne paints, even the co-solvent types.

Solvent-borne, also called **oil-based**, paints can have various combinations of organic solvents as the diluent, including aliphatics, aromatics, alcohols, ketones and white spirit.

6.3.3 Pigment or filler

Pigments are granular solids incorporated in the paint to contribute to colour. Fillers are granular solids incorporated to impart toughness, texture, give the paint special properties, or to reduce the cost of the paint. Alternatively, some paints contain dyes instead of or in combination with pigments.

Fillers are a special type of pigment that serve to thicken the film, support its structure and increase the volume of the paint. Fillers are usually cheap and inert materials such as talc, lime, clay, etc. Floor paints that will be subjected to abrasion may contain fine quartz sand as a filler. Not all paints include fillers. On the other hand, some paints contain large proportions of pigment/filler and binder.

Some pigments are toxic, such as the lead pigments that are used in lead paint. Paint manufacturers began replacing white lead pigments with the less toxic substitute before lead was banned in paint for residential use in 1978.

6.3.4 Additives

Besides the three main categories of ingredients, paint can have a wide variety of miscellaneous additives, which are usually added in small amounts, yet provide a significant effect on the product. Some examples include additives to modify surface tension, improve flow properties, improve the finished appearance, increase wet edge, improve pigment stability, impart antifreeze properties, control foaming, control skinning, etc. Other types of additives include thickeners, stabilisers, emulsifiers, texturisers, adhesion promoters, UV stabilisers, flatteners (de-glossing agents), biocides to fight bacterial growth, and the like.

Additives normally do not significantly alter the percentages of individual components in a formulation.

6.4 Timber flooring

Wood or timber flooring refers to any product manufactured from timber that is designed for use as flooring. Wooden floors are commonly choosen for their elasticity, flexibility, aesthetics and structural abilities.

Bamboo flooring is also considered to be a type of wood (timber) flooring. Some architects also use bamboo against walls for decorative purposes.

Timber floors can either be classified as **solid wood** or **engineered wood**, but we will discuss the differences below.

Do you still remember the timber section discussed in Chapter 4 of this book? Please refer back to it.

6.4.1 Solid wood

Solid hardwood floors come in a wide range of styles and dimensions with each plank made of solid wood and milled from a single piece of timber. These floors were largely used for the structural integrity and installed perpendicular to other wooden support beams commonly called joists or bearers. However and more recently, solid wood timber floors are placed directly onto concrete floors using battens.

Depending on the type, nature, grain and finishing technique employed, a natural solid wood timber floor will present a striking appearance. These timbers can be manufactured in various sizes but their application will determine the most suitable one.

Solid wood can be cut into three styles, namely **flat-sawn**, **quarter-sawn** and **rift-sawn**, with quarter-sawn and rift-sawn having the same appearance once the wood is put in place (because only one side of the plank is visible). In order to reduce cupping, grooves are usually cut along the length of the plank on the underside. Often, solid wood floors are cut with a tongue-and-groove technique for ease of installation and stability.

6.4.2 Engineered wood

Engineered wood is composed of two or more layers of wood in a plank. Normally the top layer provides the aesthetics whilst the lower layer (the core) provides the stability.

Laminate, vinyl and veneer floors can be confused with engineered wooden floors because laminate uses an image of wood on its surface whilst vinyl has a plastic formed to look like wood. Veneer floors are thin layers of wood with a core made out of a number of different composite wood products.

Engineered wood is commonly used throughout the world, whilst some countries, like America, still like to use solid wood in their timber structures.

It is often difficult to compare engineered wood to solid wood as the former can vary greatly in quality. Engineered floors are typically pre-finished with bevelled edges, which influence their appearance compared to the rougher edges of solid wood. Solid wood also has some limitations due to the following variables:

- **Moisture ingress:** We all realise the impact of moisture on the behaviour of timber (as explained in chapter 4) and this can result in the timber behaving differently under certain conditions. It is therefore important to protect the timber floor from the ingress of excessive moisture
- **Gapping:** Excessive space (gaps) between planks normally viewed longitudinally (along the length of the timber plank)
- **Crowning:** Convex curving upwards when humidity (moisture) increases
- **Cupping:** A dished or concave appearance with the height along the edges of the plank being higher than the centre.

6.4.3 Floor finishes

Polyurethane and oil are the most popular applications for wood flooring and each have differing finishing and maintenance requirements.

Some of the finishes are:

- Natural shellacs, lacquers, waxes and varnishes which are often blended with oils
- Some rural and cultural communities use cow dung to help protect the floors – these are also largely based on natural or vegetable oils
- Oils, except petroleum-based ones, have been used for thousands of years and for many years this was the most common treatment for timber floors

> Do you still remember that bitumen is considered a petroleum-based product, as discussed in chapter 3 of this book? Please refer back to it.

- **Polyurethane floors** were introduced in 1940 and have gained popularity ever since. Although there are several types, based on the different manufacturers (or brand names), the most common types are straight polyurethane and oil-modified polyurethane. Common names for these products are also urethane, lacquer or varnish

■ **Sanding** is another form of wood treatment where the floor is smoothed using either sandpaper or a floor sanding machine. Usually this is done once the floor is already in place and the sanding process is used to eliminate unevenness or scratches in the floor boards (planks). Once the sanding is completed, the wood can be stained or lacquered/varnished with several applications to improve its appearance.

6.4.4 Maintenance

There is no use spending several thousands of rands to create or prepare a beautiful timber floor, but nothing on its regular maintenance. Like with all products that are continually used, wear-and-tear will take place which will affect the appearance of the timber.

Usually proper vacuuming, sweeping or wiping with a damp cloth or mop is sufficient to maintain the cleanliness and appearance of the timber floor. It is strongly advised that cleaning agents and chemical detergents should not be used to clean the floors as they strip the oils from the wood and allow more rapid deterioration. It is important that the manufacturer's recommended cleaning products or just plain water be used when cleaning the floors. If the wood is properly prepared and kept clean, timber floors will not accumulate hidden dirt or odorous compounds.

6.4.5 Installation techniques

Timber floors can be manufactured with a variety of different installation systems. Examples of the most common techniques are:

1. **Tongue-and-groove:** The plank on one side and one end has a groove cut into it. Another plank would have a tongue (protruding wood along an edge's centre) cut into its side and end. When installing, the tongue and groove fit snugly together, thus joining or aligning the planks, and are not visible once joined. Tongue-and-groove flooring can be installed by glue-down (both engineered and solid), floating (mostly engineered), or nail-down (not recommended for most engineered wood).

2. **'Click' systems:** There are a number of patented 'click' systems available. A 'click' floor is similar to tongue-and-groove, but instead of fitting directly into the groove, the board must be angled or 'tapped' in to make the curved or barbed tongue fit into the modified groove. No adhesive is used when installing a 'click' floor, making board replacement easier. This system not only exists for engineered wood floors, but also for engineered bamboo and a small number of solid floors and it is designed to be used for floating installations. It is beneficial for the do-it-yourself market.

3. **Floor connection system:** Depending on the manufacturing technique, there are a wide range of systems available. The general principle is to have grooves on all four sides of the plank with a separate, unconnected piece that is inserted into the grooves of two planks to join them. The piece used for the connection can be made from wood, rubber or plastic. This installation system allows for different materials (i.e. wood and metal) to be installed together if they have the same connection system.

4. **Glue-down method:** Another method of installing wooden floors is to glue the wood directly onto the floor, called the glue-down method. Solid timber parquet flooring or vinyl flooring installations directly onto concrete floors use this method to secure the floor in place. The process is similar to that of laying tiles, where you spread the floor with a layer of glue or mastic using a trowel and lay the wood pieces on top of the glue, using a rubber mallet to hammer or press these into place. In order not to damage the actual timber piece, a batten or similar type of timber is placed on top of the parquet and then hammered. Once all the blocks are placed, they are often sanded to take out all unevenness and then lacquered or varnished.

Activity 1

We have tried to give some examples of timber floor applications within this chapter but given the constraints, can only discuss the basics.

1. Investigate and prepare a poster on the different timber floor applications utilising different wood types and finishes.
2. Once the above is complete, use one of the investigated topics and concentrate on preparing a second poster highlighting this specific application.

It would be an idea to create a poster competition (with prize money or some form of reward) in order to display your works and allow the rest of the class to adjudicate the best poster presentation for both the above exercises.

Activity 2

Now that you have constructed your house with the various materials as directed in the preceding chapters, it is time to finish the inside of your home. You must therefore decide how and what materials, as identified in this unit, you will be using to complete your home. As an example you may decide to have the lounge area floor constructed using a timber floor whilst the rest of the house will be tiled. All the walls must be painted and wall tiles must be placed in the bathrooms. Identify all the areas in the house where you will be using the different types of materials mentioned in this unit. Measure the surface area of each room for the application you have in mind. Once you have totalled all the areas, determine what it will cost (just for the materials) to finish your house off neatly on the inside.

Remember that some of the materials, like paint, must have more than one application on the wall and you therefore have to consider this when calculating the quantity of materials required. Also, when tiling, don't forget about the paste (glue) and grouting and when using timber floors, consider a similar approach to that of paint.

Self-evaluation 6.1

Answer the following questions:

1. What are the generic types of floor coverings typically used in buildings as finishes?
2. Describe the difference between a porcelain tile and a ceramic tile.
3. Name the two methods used to manufacture ceramic tiles.
4. What is a slush mould method as used in the manufacture of ceramic tiles?
5. Name the four most common types of natural stone tiles used in industry.
6. Use a table to describe the differences between the four most common natural stone tiles.
7. What type of rock are slate tiles formed from?
8. What is the most important ingredient of any type of paint?
9. How do we categorise binders as used in paints?
10. Why do you think lead was prohibited for use in residential paint applications?
11. What is the ingredient of paint that is restricted by environmental decree?
12. What is the purpose of the diluents in paint?
13. What are fillers and how do they function?
14. Is bamboo also a form of timber flooring?

15. What is engineered wood?
16. What are the typical limitations to be kept in mind when dealing with natural or solid wood floors?
17. Why is it important to regularly maintain timber floors?
18. Name and describe the techniques used when installing timber floors.
19. Is parquet flooring also a type of timber floor? Describe what parquet flooring is.

Answers

Self-evaluation 6.1

1. Textiles, tiles, timber and vinyl coverings.
2. Generally a ceramic tile is made of pure clay whilst a porcelain tile has a mixture of clay and other compounds.
3. Dry press method and extruded method.
4. The clay material is very wet when placed in the mould where after it is placed in an oven for drying.
5. Granite, marble, limestone and slate.
6. The learner is to prepare a table/spreadsheet which identifies the differences between the four common types of natural stone tiles.
7. Metamorphic rock.
8. The binder which imparts adhesion and influences other characteristics.
9. Binders are categorised according to the mechanisms for drying and/or curing.
10. Lead is considered harmful to both humans and the environment, i.e. it is toxic.
11. VOC's or volatile organic compounds.
12. To dissolve the polymer and adjust the viscosity of the paint.
13. Fillers thicken the paint and contribute towards the colour.
14. Bamboo is not generally considered suitable as timber flooring.
15. Engineered wood is a combination of two or more layers of wood which, when combined, provide both aesthetics and stability.
16. Moisture, gapping, crowning and cupping.
17. To maintain its appearance and strength because of wear-and-tear as well as environmental factors.
18. Tongue-and-groove, click system, floor connection system as well as glue application. Students to describe the techniques.
19. Yes, parquet is timber cut into strips or blocks and is applied to a surface by a glue application.

Index

J

joints
  bricks 204
  construction 129
  fillers and sealants 129–130
  movement 129

K

kilns
  firing bricks 193–194
  seasoning 218, *219*
kinematic viscosity test 171

L

laboratory test methods
  California bearing ratio 62
  flexural test 63
  tensile test 62
  unconfined compressive strength 62
  wet/dry test 62
lime
  cracking 62
  density 62
  pozzolanic reaction 62
  reaction with damp soil 62
  safety precautions 60
  soil suitable for treatment 63
  stabilisation 59, 61–63
  use in wet or damp conditions 62
  limestone 272

M

materials behaviour
  brittle 143
  ductile 143
  viscous 148
metals
  ferrous 229
  ferrous and non-ferrous *226*
Mines and Works Act 44
modulus of elasticity 228
moisture in soil, prevention of slopes from
  collapse 64–66
moment of inertia 228
mortar 78, *79*, 81

N

neutral axis 228
Newtons (force) 226

O

organic solvents, timber preservative 222
ozone depletion 260

P

painting
  additives 275
  binder, vehicle or resins 273–274
  diluent or solvent 274
  pigment or filler 275
particle charge test, bitumen 172
paste 78, *79*, 89, 105
penetration test, bitumen 152, 165–166
planting, slope protection 65
plasticisers *see* water-reducing agents
plastic settlement cracks 89, 90
plastic shrinkage 89, *90*
poise 167
polymer-modified binders, types added to
  bitumin 158–159

Q

quality code of practice *see* ISO 9000 series
  quality systems
quarry
  development 40–44
  dust control 43
  loading and transport 51
  rehabilitation requirements 43
  roads 42–43

R

reinforced earth 67–69
  advantages 69
  applications 68
  compaction 69
  construction methods 68–69
  facing 69
retarders 116
rip-rap, slope protection 65
road construction materials
  bitumen as basecourse 175
  bitumen as crack sealer 175
  bitumen as dust binding 175
  bitumen as fog seal 174
  bitumen as hotmix asphalt 175
  bitumen as prime coat 174
  bitumen as recycled 175
  bitumen as slurry seal 174
  bitumen as soil stabilisation 174
  bitumen as surface dressing 173–174
  bitumen as tack coat 174
  sampling *161*

S

sampling
  bitumen 160–164

www.ingramcontent.com/pod-product-compliance
Lightning Source LLC
Chambersburg PA
CBHW061804210326
41599CB00034B/6869